Expose, Excite, Ignite!

An Essential Guide to Whizz-Bang Chemistry

Carl Ahlers

Prof Bunsen Science Publishers
Melbourne ◆ Australia

Published by
Prof Bunsen Science Publishers
P O Box 7477, Geelong West, VIC, 3218, Australia
www.profbunsen.com.au

ISBN: 978-0-9870858-0-1
BISAC Category SCIENCE / Experiments & Projects

National Library of Australia Cataloguing-in-Publication entry:

Author:	Ahlers, Carl.
Title:	Expose, excite, ignite an essential guide to whizz-bang chemistry / Carl Ahlers.
ISBN:	9780987085801 (pbk.)
Notes:	Includes index.
Subjects:	Chemistry--Experiments
Dewey Number:	540.78

Trademarks

All trademarks are the property of their respective owners. The author and the Publisher are not associated with any product or vendor mentioned in this book. Nor do they endorse any product mentioned.

Cover design by Justine Elliott

Cover photo by Michael Rudolph, The Wimmera Mail-Times, Horsham. Pictured are Rod Kirkwood of Horsham College, Victoria, Australia, his students and the author.

For my parents

Hein & Trieks Ahlers

both exemplary teachers
who instilled the urge to "teach creatively" in me

"You teach some by what you say,
teach more by what you do,
but most of all you teach most by who you are"

- Author Unknown

About the Author

Carl Ahlers is a passionate science communicator and developer of commercial science teaching aids for the classroom. He is the founding director of Prof Bunsen Science located in Geelong, Australia. Carl holds a MSc in Physical Chemistry, BEd and HED and has worked as a chemist in various laboratories including military explosive development and building material science. But above all he loves to teach and has taught chemistry & physics at high school and tertiary levels. As a communicator he has many years experience in presenting science shows, special science demonstration lectures and science activities. Carl's passion for "simple, fun, engaging science" is bringing excitement and opening new opportunities to thousands of teachers and students.

Carl was born and raised in South Africa where he has delivered many popular lectures on science teaching and chemistry. In 2005 he moved to Australia with his wife and two sons.

He can be contacted on carl@profbunsen.com.au for presentation enquiries and feedback on this book.

"He who teaches, learns"
- Latin proverb

Acknowledgements

Education is about sharing skills and knowledge so students can get the maximum benefit. I have been fortunate enough to share with many talented teachers, colleagues and friends over the years who have all made a contribution - large or small - even if only adding a punch-line to a demonstration or suggesting an alternative presentation method. Great science demonstrations are like food recipes - they are handed down from one generation to the next. So, to all who have contributed, even unknowingly, I offer my sincere thanks.

I am indebted to many people in the preparation of this book. I have been privileged to sound my ideas off of Prof Jeff Bindon and Dr Cedric Smith. Thanks Jeff, for always sharing your creative ideas and taking the time to read through my text. I am also lucky to have the services of Dale Carroll from Geelong College who has edited the text and made some valuable suggestions. In my formative years, Prof Nic Basson has been a great mentor and valuable advisor. I honour him for that.

Thanks to my friend and science partner, Godwyn Morris, for her inspiration and support. Diane Spicer has done an excellent job in proofreading the text and Elzette Bester has been a great sounding board in my self-publishing effort.

Lastly, to Abrama, Richert and Evert, thanks for sharing me with my other love and providing me with the opportunity to set my creativity free!

Risk / Safety Icons used

★☆☆☆☆	No apparent risk. Take general lab precautions when using fire.
★★☆☆☆	Use of fire, hydrogen ignition or a strong base. Use general fire safety precautions. Plan demonstration well before execution.
★★★☆☆	Reactions with expanding flames or projectiles. Read through instructions twice. Plan well and try on own before demonstrating in class.
★★★★☆	Reactions with higher level of potential risk. Very high temperatures may be achieved. Plan well and read through instructions twice before doing on a small scale. Follow all instructions exactly as described.

Image credits

Page		
38, 93, 101	Sketch	Niel Venter
46, 51, 53, 62, 86, 90, 112	Sketch	Vikrant Singh
47	Cracker image	123RF stock image
49	Labware image	123RF stock image
60	Image	Sara O'Shaughnessy, Bunyip PS
64	Microchip image	123RF stock image
65	Image	Joliet Jake Szerkeszto
85	Image	NASA
90	Explosion image	Michael Rudolph, The Wimmera Mail- Times
Front & Back cover	Layout	Justine Elliott

"It's not that I'm so smart, it's just that I stay with problems longer"
Albert Einstein

Table of Contents

Disclaimer

- ✓ The author and publisher have made every reasonable effort to ensure that the experiments and activities described are safe when conducted as instructed but neither assume any liability for damages caused or injury sustained from conducting activities in this book.
- ✓ All information and material supplied are viewed and interpreted solely at the readers' risk. Readers have to develop and follow procedures for the safe performance of the activities in accordance with their local regulations and requirements.
- ✓ The author does not intend to explain all dangers known or unknown that may exist in performing any of the activities.
- ✓ All material and information are presented solely for educational and entertainment purposes.
- ✓ Responsible adults should supervise young readers who may want to undertake the activities described.

Introduction

About this book

This book is simply a collection of my favourite exothermic combustion demonstrations that I have conducted in my classroom, during science shows and at science conferences. All of the activities have great intrinsic educational value and I have used them as such in my classroom. Notwithstanding - they are all very popular with students.

I have liberally added subject content to the demonstrations so readers can have the knowledge and background information to turn a great demonstration into an educational event. The main intention of the book is to show you **how** the demonstrations are done. I tried to be as specific as possible when describing procedures so science teachers with basic chemistry skills can reproduce the activities.

When I selected activities, I simply focused on three parameters:
All activities had to be Simple, Spectacular & Safe.

- **Simple**: I know that teachers are busy and do not have the time to mix or put together elaborate concoctions. The same goes for home experimenters who want quick results. My experience is that success here will breed more success. Once you start venturing into the 'extraordinary', the feedback and success will propel you to the next level.

- **Spectacular**: Libraries and the internet provide an ample supply of books and webpages offering standard "mid-range" science content. This book offers those special spectacular activities that you require on open days and for special curriculum events. These are the ones your students are likely to remember long after they have forgotten your name . . .

- **Safe**: It had to be safe within the scope of being spectacular. Now that is a balancing act as this book is mostly about combustion! I have selected activities that do not use strong oxidants and have stuck to small volumes to keep it safe. You won't find any chlorates or perchlorates in here. Also no peroxides or concentrated acids. But you will find heaps of energy produced by icing sugar, hydrogen gas and pencil sharpeners!

I have performed and conducted the described activities so many times that I may be able to perform them blind folded. They have been tested well and I gladly back them up with the confidence that they do work as described.

This book also offers some of the practical "trade secrets" of chemistry teaching that no formal education ever passes onto new teachers. You won't find the information mentioned here in education faculties, but the truth is that chemistry teachers in the trenches engage in spectacular Whizz-Bang stuff - without the training!

It is important that teachers learn how to perform the spectacular demonstrations in a simple, safe and responsible way. Conducting demonstrations with potentially hazardous chemicals in front of an audience can be very challenging. Solid knowledge and guidance from those who have been there can fuel your confidence.

Why do we do science demonstrations?
Firstly, it is sometimes the only way in which we can link an abstract concept to the students' world - the demonstration simply acts as a bridge between the real and abstract world. Talking about "increased surface area as a factor that increases reaction rate" is abstract - but demonstrating the spectacular visual effect provides an experience to remember.

Secondly, demonstrations provide powerful motivation and make the teaching and learning of chemistry much more enjoyable. After presenting (safely) to almost half a million students, I can testify to this.

I wish you many hours of discovery and excitement with your students or in your own backyard. Let's share the joys of chemistry, safely, with everyone!

Carl Ahlers
Australia, 2011

carl@profbunsen.com.au
www.profbunsen.com.au

Everything should be made as simple as possible, but not simpler.
- Albert Einstein

Explosions, Detonation & Deflagration

Explosions are rapid violent chemical changes a substance or mix of substances undergo to produce large amounts of gas and heat, accompanied by light, sound and a high-pressure shock wave.

The general term "**explosion**" is usually used to describe the **destructive nature** of a **sudden** chemical change. In technical terms only detonating compounds should be included but as some deflagration processes also cause mechanical destruction, they are classified in popular terms as explosions too.

Deflagration is a technical term describing **subsonic combustion** that usually propagates through **thermal conductivity**. Here the expanding heat wave transfers heat to neighbouring substances, heating them up to their auto-ignition temperatures - eg. lighting a candle, bushfire, gunpowder. A process such as a dust explosion should rightly be named - dust deflagration - but carries the explosion label due to its destructive nature.

Low explosives burn or deflagrate comparatively slowly and are employed as propellants in firearms and in blasting. Well-known examples are: gunpowder, cordite, pyrotechnics and rocket fuels.

Detonation is a term describing **supersonic combustion** that propagates through **shock compression**. Here the rapidly expanding pressure wave propagates at a speed exceeding that of sound. In many explosive armaments a set-up exists that is known as an 'explosive train' or triggering sequence. A fuse is initiated by impact or time mechanism and it sets off a primary explosive causing detonation, eg. lead azide in a detonator. The resultant pressure shock wave activates the next explosive in line and finally the amplified pressure wave activates the secondary charge, eg. TNT or RDX.

High explosives decompose very rapidly in an uncontrollable blast that is labeled - detonation. They all require a detonator to be activated. Well-known examples are TNT, RDX, ANFO, dynamite and nitroglycerine.

1 Fossil Fuel Combustion

The Barking Bunny Can & Tube Cannon

If there is one request all science teachers are constantly confronted with, then it is this one:
"Please, can we blow up something ?"
Why do students ask this question? Well, most students love to experience a bang but probably because this is the way 'science' is portrayed in all 'MythBuster®' type TV programs! Here is a great idea for you to fulfill your students' wishes and teach them some valuable chemistry along the way!

The Background

Combustion or **burning** in general is a chemical reaction in which a substance reacts rapidly with **oxygen** with the production of **heat** and **light** also known as **fire**. It always requires **activation energy** (spark or flame) to get going. In carbon rich fuels it is also seen as the 'oxidation' of carbon to form its oxides - carbon dioxide and carbon monoxide. The **exothermic** nature of this reaction makes it a valuable source of energy to power our everyday lives. All fossil fuels are carbon rich and the reaction produces carbon dioxide (uuh!) and water (ooh!) as products.

Fossil fuels are the fuels man predominantly uses and includes coal, oil and natural gas or their products. These are the remains of living organisms with a high **carbon** and **hydrogen** content. (Carbon is the element of life that made up the bulk of dinosaurs and the organisms before them). These fuels include most of the things we burn to energise our world, like candle wax, coal, paraffin, turpentine, methylated spirits, lamp oil, LPG, butane, propane, methane, petrol and diesel.

In this chapter we will focus on the use of one of them - butane - that is conveniently packaged and sold at tobacconists for our use.

Figure 2.1

This is the reaction of **butane** with **oxygen**:

$$2C_4H_{10} + 13O_2 \rightarrow 8CO_2 + 10H_2O + \text{energy}$$

We will inject a quantity of butane into a Nesquik® can (the 'bunny can') and ignite it with a spark from a lighter. The spark ignites the combustion mixture causing an **increase in pressure** because:-

- the lid is still in place and the volume is fixed
- the mole gas volume change from 15 to 18 tries to increase the volume
- the exothermic (heat producing) nature of the reaction

The hot gaseous reaction products push against the can lid and walls until the pressure is high enough to blow the lid off (Figure 2.2). The bang you hear is the air pressure wave reaching your ears just like that from a popped balloon. Because the lid releases when the burning is still in progress a flash of flame can be seen.

Figure 2.2

It is incorrectly assumed by many that a bullet gathers speed as it flies toward the target. The highest speed during the flight is that of the bullet (or lid) as it leaves the muzzle ("muzzle velocity"). It would have travelled forever, if it weren't for the air resistance, gravity and in our case the fishing line that brings it to rest.

The above balanced equation is important as it gives us the ratios in which reactants react also known as "the ideal stoichiometric combustion ratio".

It also tells us that there is no use in adding more fuel, if oxygen is not supplied too. The equation states that two moles of butane reacts with thirteen moles of oxygen. Or 2/13 = 0.15 moles butane reacts with one mole of oxygen.

As the Ideal Gas Law is closely obeyed by nearly all gases at room temperature and standard pressure, the ratio of 0.15 : 1 should thus apply to the *volumes* of the gases too.

The amount of **oxygen** in our 500 ml can is set. Air contains 21% oxygen therefore:

21% of the volume of the can: $0.21 \times 500\ cm^3 = 105\ cm^3$ oxygen
therefore requires: $105 \times 0.15 = 15.8\ cm^3$ butane.
Volume ratio: butane volume : air volume
15.8 : 500
1 : 32

So, for butane you may simply divide your can's volume by 32 to calculate the volume of butane required. Even if you work in fluid ounces. Of course, if you use other fuel types the ratio will change.

For **propane (C_3H_8)**

$$C_3H_8 + 5O_2 \rightarrow 3CO_2 + 4H_2O + \text{energy}$$

we have 0.20 moles propane reacting with 1 mole oxygen and following the same reasoning as above, the propane : air volume ratio becomes **1 : 24**.

By the way: The ideal fuel-air volume ratio for petrol (gasoline) in motor car engines is 1 : 14.7 but this will change depending on the type of fuel.

A. The Barking Bunny[1] Can

[★★☆☆☆]

Now let's get on with the practical side. We will use a simple approach so that the device can be made in minutes. We are after a demonstration that is reliable, safe and easy to produce.

Figure 2.3

Safety / Risk Assessment

★ The combustion process produces heat and expanding

[1] The bunny refers to the trade marked and universally well-known Nesquik® bunny.

gas, so expect a loud noise and a fast moving lid. The lid should be tied down with fishing line as instructed below. You should wear eye and ear protection and caution the students to protect their ears. Do not launch any projectile at glass objects, animals or people.

★ Furthermore - take general precautions when drilling metal and be on the lookout for burrs and sharp edges.

What you will need

- One 250 g Nesquik® can (or similar can with a metal press-lid)
- Measuring cylinder / cup
- Fire lighter with round shaft
- Drill and drill bit with diameter same as fire lighter shaft
- Butane refill can (used for refilling of gas lighters, at tobacconist shop)
- Plastic syringe (20 mL or 0.7 oz) - no needle
- Fishing line (strong!) approx. 1 m (3.3')
- Cable tie that will fit around the can.
- Poking tool, awl or nail
- File or sand paper

Figure 2.4

Here's how

1. Clean a Nesquik® can (or similar metal can).

2. Measure and record the volume of the can using water and a measuring cylinder / cup.

3. Poke or hammer a hole into the centre of the base with the poking tool or nail (Figure 2.5).

4. Enlarge this hole to the diameter of the gas lighter shaft with the hand drill (Figure 2.6). The lighter shaft should fit tightly through this hole (Figure 2.7). This can be done with long nose (pointy) pliers too.

Figure 2.5

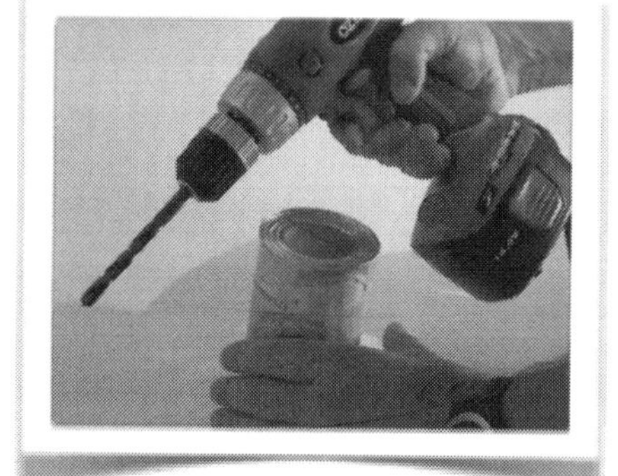

Figure 2.6

Simply rotate the pliers in the hole until it is the right size.

5. Remove all outside burrs from the hole with a fine file or sand paper. Careful!

6. Cut a length of strong fishing line - about 1 m or 3.3' Poke or drill a small hole in the **rim of the metal lid** and tie the fishing line securely to the lid (Figure 2.8).

7. Position the cable tie close to the bottom end of the can but do not pull it tight yet (Figure 2.9).

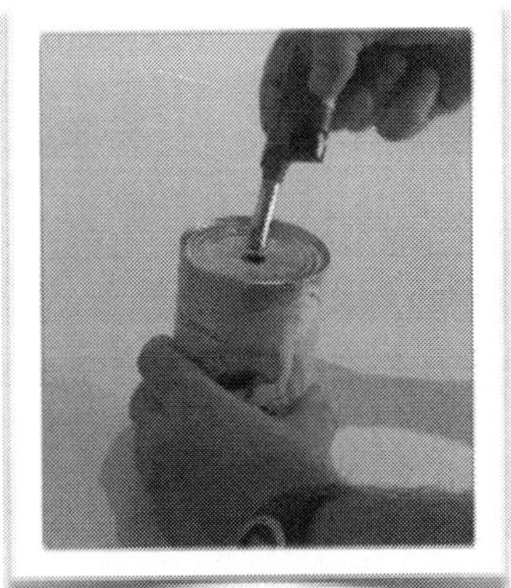

Figure 2.7

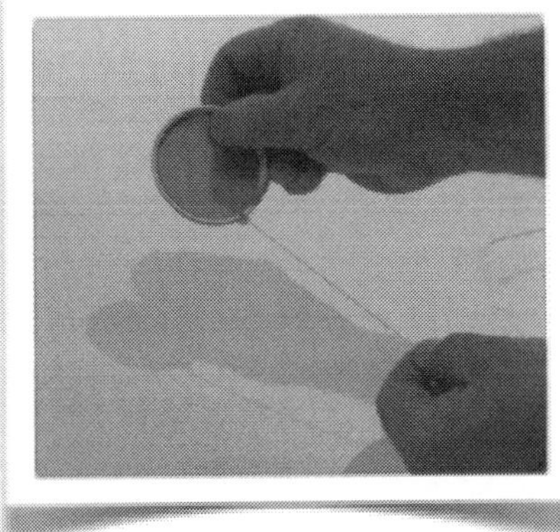

Figure 2.8

Figure 2.9

8. Now, tie the other end of the fishing line securely to the cable tie and then pull the cable tie tight.

Operating the Bunny Can

1. Calculate the amount of butane gas required for combustion:

   Volume of can ÷ 32 = x mL butane.

2. Push the metal lid securely onto the can.

3. Hold the butane can upright and position the plastic syringe tip on the butane can nozzle and simply press the syringe down (Figure 2.10). The syringe plunger will shoot up. Fill to x ml.

4. Now inject the butane **rapidly** into the can to promote good mixing (Figure 2.11). Then cover the hole with your finger. The gas will slowly diffuse throughout the can - allow a few seconds. (To improve the

mixing, see 'Adding a mixer' on page 21).

5. Insert the gas lighter (figure 2.12), do the countdown and pull the trigger as **sharply** as possible (because the lighter may flood the tip with extra fuel) . . . Fire in the Hole!
   Safety: Aim away or well above your spectators.

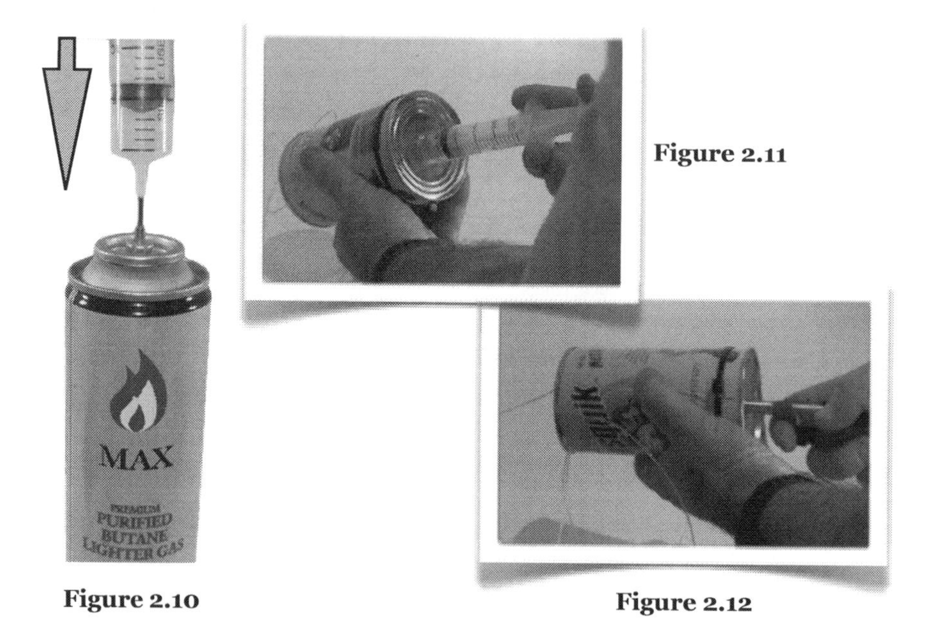

Figure 2.10

Figure 2.11

Figure 2.12

6. If it does not fire, withdraw the lighter, re-insert and try again. If still unsuccessful, open the lid, blow into the can to refresh the air supply and then repeat the procedure. Always refresh the air before firing again.

Teaching Extensions

Here are a few ideas to apply in your classroom or backyard research:

1. Why did the lid fly off and (luckily) not the lighter?
   Start with the definition of pressure (P): $P = F/A$ where F is the force and A the surface area. Focus on the surface areas of the lid and lighter and remember that the pressure inside the can is equal in all directions.

2. Can your students calculate the ideal stoichiometric combustion ratio of methane (CH_4) to air? How about other fuels?
   They first have to balance the equation

$$CH_4 + O_2 \rightarrow CO_2 + H_2O + \text{energy}$$

   to get the molar ratio. The gas volume ratio is 1 : 9.4

3. Replace the butane with hair spray or deodorant spray. You will now have a mixture of butane and propane plus some impurities you did not have before.

4. Over-fueling will create a more visible flame.
This is pure edutainment! It is important (and educational) to emphasize the result of an over or under supply of reactants.
Refer students to the balanced equation. When using an oversupply of butane, the gas mixture usually does not ignite due to insufficient oxygen. But when the flame emerges from the can, it finds new oxygen in the outside air and burns brightly.

Try it: Inject double the volume you used before and you will find it will be hard to ignite. But, when the lighter is held outside the opening, the gas ignites and produces a rocketlike fire plume (Figure 2.13).

Figure 2.13

Combustion reactions perform at their **peak** when combustion ratios are **stoichiometrically ideal** as per the balanced equations. But it is important to realise that fuel gases can react over much wider ranges than that predicted by the balanced equations. Not withstanding this, this is still a great way to make students realise the importance of balancing equations to find the ideal stoichiometric ratios!

Table 2.1 lists the explosive or flammable range of gas/air mixtures. Below the indicated value, the mixture is too lean to burn and above the upper explosive limit the mixture is too rich to burn. The limits indicated are for gas and air at 20°C and atmospheric pressure[2].

Let's take butane as an example:
We injected 15.7 / 500 x 100 = 3.13% fuel into the can which is fine as butane can ignite between 1.8 to 8.4%

---

[2] The Engineering Toolbox, www.engineeringtoolbox.com
An invaluable resource.

Fuel Gas	Flammable Range (% by volume)
Hydrogen	4 - 75
Methane	5.0 - 15.0
Butane	1.8 - 8.4
Propane	2.1 - 10.1

Table 2.1

B. The Tube Cannon

[★★☆☆☆]

You may want to extend the Bunny Can demonstration by launching the lid or a plastic cup from a Pringles® tube. This is loud but fortunately the projectiles are safe.

What you will need

- all items listed for the Bunny Can project except the cable tie and fishing line
- a large Pringles® tube
- plastic cup ~ 100 mL (3.4 oz) that will fit the tube - or almost
- scissors

Here's how

1. To make the hole for the gas lighter, hammer a nail through the centre of the tube's base or use a poking tool.

2. Enlarge this hole to the diameter of the gas lighter shaft with a hand drill or pointy pliers. The lighter shaft should fit tightly through the hole.

3. There is no need here to tie the plastic lid down due to its small mass.

4. Measure the volume of the Pringles® tube. It should be close to 950 mL (32 oz). Divide this by 32 if you use butane gas:

   950/32 = 30 mL (1 oz)

Operating the cannon

1. Place the plastic lid onto the Pringles® tube.

Figure 2.14

2. Collect 30 mL (1 oz) of butane gas with the syringe and inject into the tube.

3. Cover the hole with your finger and wait a few seconds.

4. Remove your finger, insert the lighter and pull the trigger. "Fire in the hole!"
   Safety: Never aim at fragile objects or people. Warn observers to protect their ears.

Launching a plastic cup

Trim the rim of the cup with scissors so it fits tightly into the tube (Figure 2.15). Proceed as before but with less gas to make up for the slightly reduced tube volume.

Figure 2.15

Adding a 'mixer'

A simple non-flammable mixer can be used in the combustion chamber to facilitate better mixing (diffusion) of the gases. Tear ¼ page from a newspaper, roll it into a tight ball and wrap with a few layers of aluminium foil. It should be no larger than 50% of the diameter of the tube. Place the mixer in the tube, inject the fuel and shake the tube & mixer while sealing the hole with your finger. Ignite the gas to see if you have better combustion.

Key Terms

Combustion, fossil fuels, stoichiometry, balancing of chemical equations, Ideal Gas Law, activation energy, limiting reagents, exothermic reactions

Acknowledgement

I am indebted to Prof Jeff Bindon of the University of KwaZulu-Natal, Durban, South Africa for his valuable ideas & comments. He is the father of the Micro Steam Car Challenge, a great fossil fuel energy challenge. These cars (Figure 2.16) are made from 100% recycled items and school students have achieved efficiencies of up to 5 km on only 20 mL of methylated spirits (That is 3.1 miles on 0.7 fl oz)!

Figure 2.16

Appendix A:
Butane as lighter gas

n-Butane or n-C_4H_{10} (bp. 0.5°C, 32.9°F) is a gas at room temperature exerting a vapour pressure of 2.4 atm. In the canister and fire lighters the fuel is stored as a liquid under pressure. By squeezing the trigger, the pressure is reduced and the liquid 'boils' to a vapour and this is expelled through the orifice. Gas storage under pressure on an airline can cause rupture if sudden depressurization occurs - hence the warning not to take lighters and aerosols along.
Why not use propane in fire lighters?
Propane exerts a vapour pressure of approximately 9 atm. at room temperature (25°C, 77°F). This is much higher than an inexpensive hand held plastic lighter can safely handle!

2 Hydrogen from Kitchen Stuff

Two ingredients from the kitchen will prove that kitchen science can be a hot topic!

Many years ago my life as a chemist started out with this amazing 'kitchen science' chemistry. I spent hours preparing hydrogen filled balloons and polluting the air with colourful home-made weather balloons. I still recall the backyard 'lab' with endless rolls of pocket money aluminium foil.

We did many things with our balloons . . . . some acted as moving targets for our air gun and some took scientific and secret messages over long distances. I even had a map in my room indicating the areas from which phone calls were received as people phoned in after collecting the tags attached to the balloons. On windless days we launched gliders from the rising balloons using the ice switch method. All in all – we had lots of fun and even though Science Fairs didn't exist at the time, we performed some "research" and even had controls built into our systems. Let's call this "informal science".

The Background

Hydrogen is the most abundant and lightest element in the universe, it is colourless, odourless and a non-toxic gas consisting of H_2 molecules. The density of hydrogen (0.0899 g/l) is less than that of air (which is a mixture of different gasses) and therefore a hydrogen balloon rises in air.

Discovered by Henry Cavendish in 1766, hydrogen was originally named "inflammable air" as it combusted when mixed with air. Something we will pursue in this chapter.

There are many ways of **preparing** hydrogen gas in the laboratory. The best known procedure being the reaction of an acid (usually hydrochloric acid) with a metal (usually zinc metal). The chemical reaction being:

$$2HCl\ (aq) + Zn\ (s) \rightarrow H_2\ (g) + ZnCl_2\ (aq)$$

Another method is that of electrolysis – the breaking up of water into hydrogen and oxygen when an electrical current is passed through it.
In this chapter we will describe a less known method which has certain advantages: The reacting reagents are less hazardous to handle than acid and readily available from the supermarket. Furthermore, it does not require batteries and special collection techniques as with the electrolysis.

Aluminium foil (Al) is a well-known product in the kitchen and is one of the rare 'pure' chemical elements we encounter in our daily lives. Because of its reactivity, pure aluminium instantly gets covered in a thin outer transparent oxide layer known as corundum (aluminium oxide).

Caustic soda, known in the lab as **sodium hydroxide** (NaOH) or traditionally as **lye**, is found in the laundry section of the supermarket as it is used to make soap or used as a drain cleaner. It is one of only a few chemicals that will penetrate the strong protective aluminium oxide layer to get to the reactive aluminium.

The chemical reaction for the preparation is:

$$2Al\ (s) + 2NaOH\ (aq) + 2H_2O\ (l) \rightarrow 2NaAlO_2\ (aq) + 3H_2\ (g) + \text{energy}$$

Sodium aluminate (black compound), hydrogen gas and heat are the reaction products.

Hydrogen and Oxygen

Hydrogen gas combines explosively with air oxygen in concentrations ranging from 4 to 75 % hydrogen by volume. It requires a form of activation energy to get going. The reaction produces water in an exothermic reaction:

$$2H_2\ (g) + O_2\ (g) \rightarrow 2H_2O\ (g) + 572\ kJ$$

For every milliliter of water produced, 15,900 J of energy is released, in the form of sound, heat, light and mechanical energy. The rapid release of a considerable amount of energy causes the pocket of water vapour formed to expand rapidly, resulting in a loud "Bang". It is also known as the oxidation of hydrogen to hydrogen oxide (water!).
When igniting a hydrogen filled balloon with a candle flame, the flame first pops the rubber and then provides the activation energy needed to

trigger the reaction. The hydrogen from the balloon reacts with the 21% oxygen in the atmosphere.

Figure 3.1

More on Hydrogen . . .

- ★ Hydrogen burns in pure oxygen producing temperatures as high as 2888°C (5230°F). The operation of the oxyhydrogen gas welding torch is based on this reaction
- ★ The Sun converts six hundred million tons of hydrogen per second into helium
- ★ Hydrogen will play a much bigger role in our futures once we manage to increase the efficiency of solar energy conversions to produce hydrogen from water
- ★ Hydrogen is used as rocket fuel in the NASA Space Shuttle program
- ★ The destruction of the Hindenburg airship, which 'deflagrated' on 6 May 1937 at Lakehurst, New Jersey in the USA, was an infamous example of the flammable properties of hydrogen gas. The cause of the fire is debated and most of the visible fires were attributed to the ship's framework. Due to hydrogen's buoyancy in air, its flames ascend rapidly and cause less damage than hydrocarbon fires. Two-thirds of the Hindenburg passengers actually survived the hydrogen fire.

A. Preparing Hydrogen Gas
[★★☆☆☆]

Safety / Risk assessment

★ Follow standard safety procedures when handling chemicals
★ Work in a well ventilated area
★ Wear gloves and eye protection throughout the preparation as well as ear protection when igniting hydrogen
★ Wear a protective lab coat or apron
★ No matches, sparks or flames should be brought close to the preparation area

Chemical Safety

Sodium hydroxide (NaOH) also known as caustic soda or lye is corrosive, destroys clothes and causes injury to the skin. Treat with the greatest respect and wash your hands under running water immediately after handling this chemical. Use protective gloves.

Risk phrases: R35 Causes severe burns
Safety phrase: S2/26/37/39 Keep out of reach of children; In case of contact with eyes, rinse immediately with plenty of water and seek medical advice; Wear suitable gloves and eye/face protection.

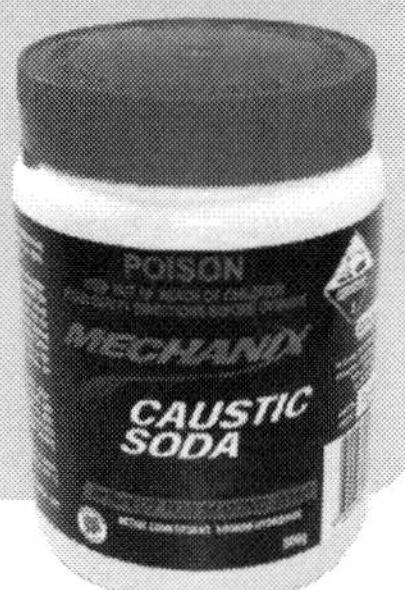

Figure 3.2

What you will need

Chemicals

- Sodium hydroxide (Caustic Soda): NaOH
  There is no need to use lab chemicals. Most supermarkets will stock this in their laundry section as it is used as a drain cleaner
- Aluminium kitchen foil or aluminium pie pans. Any thickness and roll size will do

Equipment

- Balloons approximately 30 cm in diameter when inflated. It is better to use helium quality balloons if you plan to have the gas stored for an hour or two
- Balloon clips from a party supply shop with pieces of string attached
- Clear glass bottle with narrow neck. Wine bottles are just perfect for this job. Do not use a plastic bottle

Figure 3.3

- Water cooler for cooling the bottle. A beaker or cut-off plastic milk container will do. Check that the glass bottle fits into it
- Plastic funnel (to fit bottle)
- Teaspoon
- Thin candle taped to a dowel stick
- Ear protection, goggles, lab coat & gloves

Here's how

1. Fill the glass bottle up to one third of its volume with water. Do NOT use a plastic bottle!

2. Stand the bottle in the container of water to cool the exothermic process and protect the bottle from cracking (Figure 3.4). The container water level should be at least the same as the level inside the bottle.

3. Safety glasses and gloves on! Carefully add **three heaped teaspoons** of caustic soda to the water in the bottle using the funnel (Figure 3.5). Be sure to close the caustic soda container as caustic soda is hygroscopic (absorbs moisture from the atmosphere). Do **not** seal the glass bottle.

Figure 3.4

Figure 3.5

4. Swirl the bottle gently until all the solids have dissolved. Take care not to slosh any solution from the bottle. *[Note that the solution will get hot as the dissolution of sodium hydroxide is an exothermic process].*

5. Tear **three sheets of aluminium foil**, each approximately 20 cm x 30 cm (8" x 12"). Tear each sheet in half and roll into a cylinder so it will easily fit through the mouth of the bottle (Figure 3.6). Pop them into the bottle.

Figure 3.6

Three sheets should be sufficient to launch one large balloon.

6. Inflate a balloon once to stretch it, let the air out and pull the balloon over the mouth of the bottle, while someone else holds the bottle steady. Gently swirl the foil and solution a few times and place the bottle back in the water container.

7. The balloon will slowly fill up with hydrogen gas (H_2) (Figure 3.7). Draw the students attention to the exothermic process, the colour change of the solution (sodium aluminate) and the formation of gas bubbles.

Figure 3.7

Safety: It is important that the caustic soda solution should not get too hot as this will produce water vapour in the balloon that will add unnecessary weight to the balloon and prevent it from floating. The reaction rate has to be slowed down. Here are some recommendations:

- *Do not add more than 3 heaped teaspoons of caustic soda*
- *Do not add more than 3 rolled aluminium sheets at a time*
- *Cool the reaction mixture in cold water using a beaker or trough*
- *Let the bottle and its contents cool before filling the next balloon*

8. When the balloon is filled to its maximum capacity or if the reaction in the bottle ceases, gently remove the balloon from the bottle while someone else holds the bottle steady.
   Take care - the bottle will be very hot.

9. Secure the balloon clip to the balloon base and check that no gas can escape.
   To fill another balloon: Wait a few minutes for the solution to cool, add more caustic soda and repeat the procedure.

   The inflated balloon should rise - if not, then
   - ★ the balloon might contain too much condensed water. The reaction proceeded too quickly – repeat the process by replacing the cooling water and reducing the concentration of the caustic soda
   - ★ the balloon might not be filled to its full capacity. Add more reagents next time

Ask students what the identity of the gas is and invariably they will reply "helium". So this may be the moment to demonstrate "helium's" inertness by making it react violently with atmospheric oxygen.

There are many things to do with hydrogen balloons. Turn to the next chapter to see what your options are, or simply explode them as described in the next activity.

Making the hydrogen accessible

Here is a novel way to deliver the prepared hydrogen "on demand". Source an air tool known as an **air pistol blowgun** at a tool or hardware shop. They are used for cleaning equipment with compressed air and only cost a few dollars. You will need to purchase a **clip-on connector** also so the balloon can fit onto the blowgun (Figure 3.8).

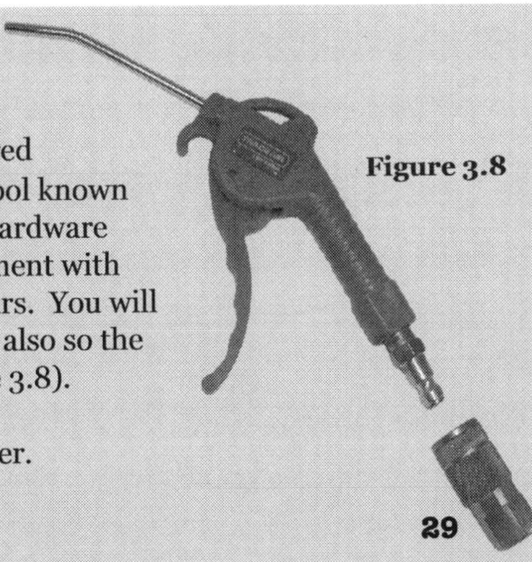

Figure 3.8

We will use this device in the next chapter.

B. Exploding a Hydrogen Balloon
[★★★☆☆]

This is the fun part that most students love, but also the more hazardous part. Do plan this carefully. The biggest potential hazard is the possibility of igniting other chemicals or combustible substances.
The explosion can be **loud** but **no solid shrapnel pieces** are produced.

Here's how

1. Tie the hydrogen balloon's string to a table, chair or other structure away from combustible material (Figure 3.9).

2. Tape a thin candle securely to a wooden dowel. This will allow you to ignite the balloon from a safe distance (Figure 3.10).

3. Light the candle.
   Safety: Warn all bystanders to stand off (at least 5 m) and to protect their ears. Wear safety goggles and ear protection. Have a fire extinguisher on hand.

Figure 3.9

4. Use the lit candle at arm's length to ignite the bottom of the balloon . . . BANG !!!!

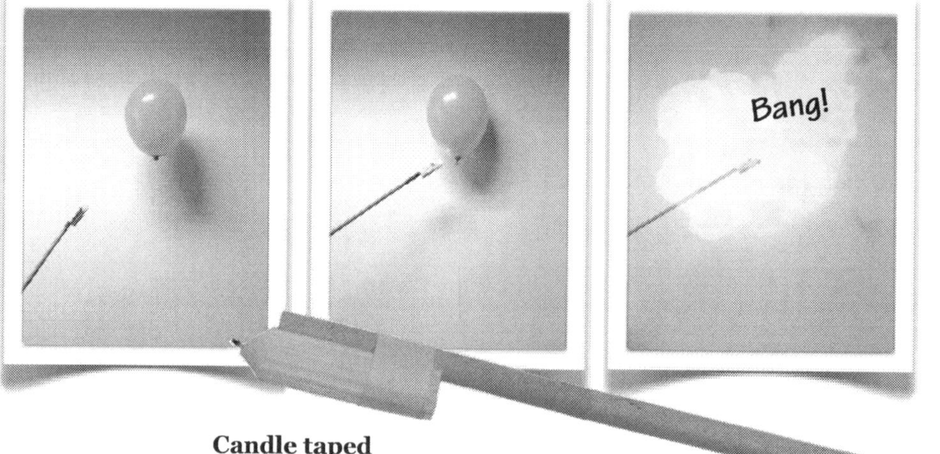

Candle taped to dowel

Figure 3.10

Disposal & Clean-up

Do not store the caustic soda solution. Dispose of it and any residue from the reaction down a toilet – this does not pose a hazard as caustic soda is the active ingredient in most drain cleaners. Rinse the bottle with water.

Teaching Extensions

Here are a few suggestions to challenge interested students:

1. Looking at the H_2 preparation,

$$2Al\ (s) + 2NaOH\ (aq) + 2H_2O \rightarrow 2NaAlO_2\ (aq) + 3H_2 + \text{energy}$$

   - ✓ Where did the $NaAlO_2$ go?
   - ✓ How would the addition of finely cut aluminium foil change the rate of the chemical reaction?

2. Why does the filled balloon shrink after a few hours, so much so that it no longer floats?
   Students can investigate 'effusion'. According to Graham's Law the rate at which gases effuse is dependent on their molecular weight. Gases with a lower molecular weight effuse more quickly than gases with a higher molecular weight. Compare the rate of effusion of two balloons filled with He and H_2.

3. What is an exothermic reaction?
   Can students identify the two exothermic reactions here?
   Water & caustic soda and caustic soda & aluminium.

4. Compare the hydrogen and butane combustion reactions:

$$2H_2 + O_2 \rightarrow 2H_2O + \text{energy}$$

$$2C_4H_{10} + 13O_2 \rightarrow 8CO_2 + 10H_2O + \text{energy}$$

   - ✓ What impact will both reactions have on issues such as global warming and our sustainable environment?
   - ✓ Emphasize the need for oxygen in both combustions. Note the molar ratios of fuel to oxygen.
   - ✓ Where did the H_2O go when the H_2 balloon was ignited?

Key Terms

Combustion, balancing of chemical equations, effusion, activation energy, exothermic reactions, hydrogen gas, sustainable energy

3 Hot Hydrogen Applications

Hydrogen is an important gas as it is the most abundant in the universe (75%), has the highest energy content per unit weight of all fuels and produces only water and energy when reacting with oxygen.

It is also easy and cheap to prepare (see previous chapter) and can produce a number of safe Whizz-Bang demonstrations with no nasty residues or flying pieces that may compromise students' safety.

All in all hydrogen is a very useful chemical for the resourceful teacher.

So what can we do with a number of hydrogen filled balloons?
"Let's celebrate the joys of chemistry in a big way!"

We will focus in this chapter on two of hydrogen's most useful properties and find some fun classroom applications:

★ Hydrogen has a low density and a lifting power 7% greater than Helium. It's density is only 0.090 g/L (Air is 1.205g /L at 20°C).

★ Hydrogen produces a highly exothermic reaction when it combusts with oxygen to form water and lots of energy.

A. Floating Hydrogen on Air

Balloons and airships are lighter-than-air (LTA), and fly because they are buoyant: the total weight of the craft is less than the weight of the air it displaces. The Greek philosopher Archimedes (287 BC – 212 B.C.) was the first to establish the basic principle of buoyancy.

Buoyancy is the **upwards thrust** acting on a body immersed in a fluid (air or liquid). It is equal to the **weight** of the fluid displaced.

$$\text{Buoyancy} = \text{Upthrust} = (mg)_{\text{fluid}} = (\rho V g)_{\text{fluid}}$$

Two equal sized balloons (of 14 litres) floating in air filled with hydrogen and helium, will experience **the same upthrust** (they displace the same weight of air):

$$\text{Weight}_{\text{AIR}} = mg = \rho Vg = (1.205\ \text{g/L})(14\text{L})(9.8\ \text{m/s}^2) = 165.3\ \text{N}$$

But as the two gases have different densities, their **weights will be different**:

$$\text{Weight}_{\text{H2}} = mg = \rho Vg = (0.090\ \text{g/L})(14\text{L})(9.8\ \text{m/s}^2) = 12.3\ \text{N}$$

$$\text{Weight}_{\text{He}} = mg = \rho Vg = (0.179\ \text{g/L})(14\text{L})(9.8\ \text{m/s}^2) = 24.6\ \text{N}$$

(We assume that the rubber balloons are of the same weight and thus are not included in the calculation. Densities: H_2 0.090 g/L and He 0.179 g/L)

So what difference does this make in lifting power (resultant force)?

Lifting power H_2: 12.3 - 165.3 = -153.0 N

He: 24.6 - 165.3 = -140.7 N

(the negative signs indicate that it is an upwards force)

153.0 - 140.7 / 165.3 x 100 = 7.44%

This gives hydrogen a 7.4% lifting power advantage over helium.

Safety

- ★ No flames or heat should be allowed close to the balloons
- ★ Never inhale hydrogen gas to demonstrate the "Donald Duck" voice effect. Use only helium for this demonstration
- ★ Check with the local aviation authority if permission is required to launch balloons

1. Balloons with Messages
[★☆☆☆☆]

A message is attached to the H_2 balloon that will faithfully float and fall with it as the balloon pops. If someone finds it and takes the trouble to call, then you will know how the winds have blown on the launch day. You may want to place a map and pins in the classroom to record the locations of the balloons found.

What you will need

- Balloons filled with hydrogen. You could fill them with lab grade H_2 or He (but that would be so boring!), or prepare your own as indicated in the previous chapter
- Masking tape
- Red or yellow cardboard
- Pen, scissors

Here's how

1. A red or yellow note (for maximum visibility) (Fig. 4.1) is attached to the balloon's string with masking tape (Fig. 4.2).

Figure 4.1

2. If the balloon does not float with the attached note, then increase either its gas volume, remove some excess rubber from the balloon's lip or cut the note smaller.

3. Check the forecast wind direction and strength with the local weather bureau. Launch the balloon and make observations as it soars (Figure 4.2).

4. From phone calls and visual observations students can draft a table with data:
   - How many seconds pass until the balloon is out of sight?
   - Does this period change during the day? (cool morning / hot afternoon)
   - What is the direction of upper air currents?

- The distance a balloon is found from the launch site.

Teaching Extension

Students might be surprised to learn that all soaring balloons return . . . very soon . . . popped! No exception. Why?

Figure 4.2

As elevation from sea level increases there are exponentially fewer and fewer air molecules. The air gets thinner (air pressure falls) the higher the balloon goes. Most people know that additional oxygen is required in the 'thin' air of Mount Everest at an altitude of 8,848 meters (29,030 feet).

There are some remarkable facts about air pressure:
The air pressure at an altitude of 10 km (32,810 ft) is only 26.2% of that at sea level. No wonder commercial airliners fly at this altitude to reduce air friction. But in exchange they have to supply air under pressure to the cabin.

At an altitude of 30 km (98,430 ft) the air pressure drops to only 1.1% of the pressure at sea level. This confirms that 98.9% of the earth's atmosphere is contained 30 km from sea level. This is assumed by many scientists as the "edge" of our atmosphere. So thin, so fragile!

Altitude	Pressure % of 1 atm	Balloon Volume dm^3 or L
0 m	100 %	14.14
3,000 m	69.2 %	20.43
5,000 m	53.4 %	26.50
10,000 m	26.2 %	54.00
30,000 m	1.1 %	1,285.00

Table 4.1

Since $P \propto {}^{1}/_{V}$ when T is constant (Boyle's Gas Law), the balloon's volume will increase dramatically as the pressure drops (Table 4.1) and since the rubber's stretch is limited, it pops!

In a simple simulation, you can place a small inflated balloon in a vacuum flask or desiccator and suck the air from the flask with a vacuum pump. The photos in Figure 4.3 tell the story:

Figure 4.3

2. Balloons with Gliders

[★☆☆☆☆]

Release a light glider from a balloon with a novel ice-switch time mechanism.

What you will need

- Balloons filled with hydrogen. You could fill them with lab grade H_2 or He or prepare your own.
- Plastic drinking straw
- String, masking tape & scissors
- Paper plane or cheap toy glider
- Freezer
- Impact heat sealer or hot glue gun

Here's how

1. Cut a plastic drinking straw into about 3 cm (1.2”) lengths. Seal one end of each using a kitchen heat sealer or you could use a hot glue gun. Check that it is water tight.

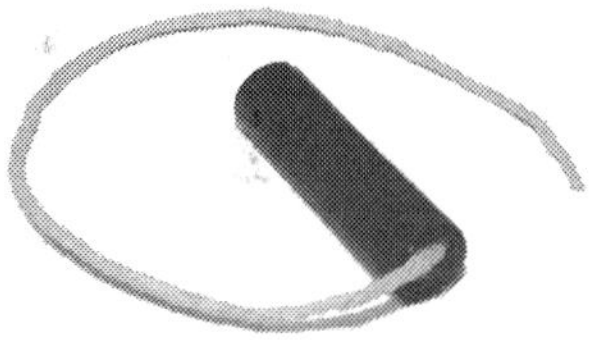

Figure 4.4

2. To prepare an ice-switch: Insert a piece of string into a plastic drinking straw with a length protruding from the straw and fill with water (Figure 4.4). Place it in an upright position in the freezer and leave for approximately 30 minutes to allow the water to freeze.

3. Remove from the freezer, note the time. Let it hang suspended on the string outside in the sun (the same conditions that will exist under the floating balloon).
   Note how long it takes for the ice to slip out (Period A).
   Repeat the experiment several times to obtain a more accurate value.

Figure 4.5

4. Decide on the time lapse after which the glider should be released from the balloon by the ice switch, e.g. 1 minute after launching (Period B).

5. Determine the waiting time before the balloon should be launched (Period A minus Period B).
   Example:
   Period A : from freezer till ice slips out : 7 minutes.
   Period B : from launching to releasing of glider : 1 minute
   Period A - B = time to wait before launching the balloon: 6 minutes.

6. Fill a balloon with hydrogen gas, prepare an ice-switch and tie the ice-switch string to the balloon. Tape the ice-switch straw to the glider with masking tape (Figure 4.5).

7. Launch the balloon at the calculated time, keeping your bicycle at hand to retrieve the glider when it lands. Enjoy!

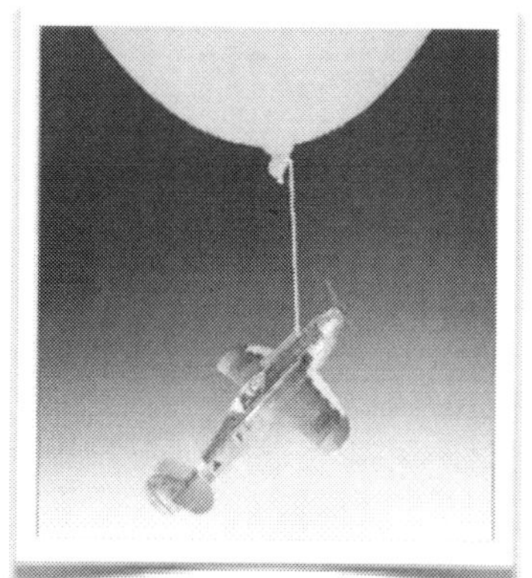

Figure 4.6

B. Making Hydrogen go Bang!

There are many ways in which hydrogen's **exothermic reactivity** can be visually demonstrated. Activities range from the meek and mild hydrogen pop in a test tube to unsafe combustion in glassware.

I once saw a teacher filling a glass milk bottle with hydrogen gas. He wrapped it in a towel (just in case . . .) and ignited it. He was lucky - the bottle did not shatter. Filling glass containers with hydrogen and igniting the gas is something one should never attempt. Please steer clear from this activity.

Safety / Risk Assessment

★ Hydrogen is not flammable on its own. It requires oxygen to combust but this can happen over quite a wide range: 7 to 75% in hydrogen
★ Stick to the instructions and quantities as advised.
★ No matches, sparks or flames should be brought close to the hydrogen balloon storage area.
★ Keep all flammable and combustible material away from the working area.
★ Wear eye and ear protection as well as a lab coat.

1. Make a Tube Whistle & Jump

[★★★☆☆]

This demonstration can be performed in a classroom but is better performed in an area with more head space. The tube can easily fly to 6 metres (20') if all goes well.

Warning: In a room the tube may dent the ceiling and be deflected back.

What you will need

- Balloons filled with **hydrogen** gas. You could fill them with lab grade H_2 or prepare your own as indicated in the previous chapter.
  It is convenient to have these balloons sealed off with balloon clips
- Tall Pringles® Tube
- Small nail (1 to 2 mm diameter, (1/16")) and hammer
- Sticky stuff (BluTack®)
- Fire lighter (long stemmed)
- Ear protection, goggles, lab coat

Here's how

1. Empty the tube (yummy!) and wipe the inside with a cloth. Hammer a nail through the center of the metal base of the tube. It should be between 1 and 2 mm (1/16") in diameter.

2. Poke a number of holes around the circumference of the tube through the cardboard, 5 mm (0.2") from the edge of the open end. These are the air inlet holes - the size is not important. (Figure 4.7)

3. Hold the tube vertical, with the metal base at the top. Remove the lid and cover the small hole at the top with a small blob of sticky stuff (BluTack®).

4. Take the hydrogen balloon, remove the balloon clip and slowly, over one minute, discharge all of the the H_2 gas from it into the tube (Figure 4.7). The tube should be held vertical. The dense air will be displaced with less dense hydrogen gas. Or connect the balloon to an air blowgun, insert the nozzle into the tube and squeeze the trigger (see previous chapter).

Figure 4.7

5. Keep the tube upright (metal base with BluTack® at top). Replace the tube's lid and stand it upright on a solid wooden table. *Check that there is nothing above the tube, eg. lights, ceiling fans or (even worse) a video projector.*

Safety: Caution the audience to cover their ears as there is a small chance of immediate combustion when igniting the H_2. Make the audience stand off at least 5 metres. Check that you have protected your ears and eyes.

6. Steady the tube with one hand on the side of the tube and remove the BluTack® (roll it off). Now light the H_2 with a **long stem** lighter so your hand isn't close by if the tube does explode. You have **a few seconds** to light the flame when opened, so take your time . . .
   You should hear a light 'pop' sound when the gas ignites and if you dim the lights, should see a yellow flame burning at the hole (Figure 4.8).

Warning: If you have not displaced all of the air inside the tube, there is a chance that the tube may explode when you light it. It is a frighteningly loud experience but won't cause any harm to anyone. More of a nuisance to refill a balloon and re-do the demo.

7. Once the flame is lit, step back and indicate to the students the yellow colour of the flame that is a characteristic flame colour of sodium. (You used sodium hydroxide in the preparation of the gas). It may take up to 3 minutes before the tube explodes. And now for the surprise! After a while the flame and tube will start to **resonate**! First as a faint whining sound and then picking up in intensity and volume, delivering a great explosive crescendo! We have a singing, exploding tube. Enjoy!

Figure 4.8

What is happening here?

Why the delay before the explosion?

The escaping hydrogen gas spews through the hole due to the column of very buoyant hydrogen gas below it. It mixes with air on the outside, ignites and combusts. The flame does not drop back into the tube because of the gas pressure from within. *The flame is actually suspended above the metal surface.* During the combustion process water forms and this can sometimes be detected as tiny water droplets on the metal surrounding the flame. The combustion uses up the H_2 gas so air is pushed in through the bottom holes in the cardboard. The air is more dense than H_2 and soon the inner tube's gas mixture can't suspend the flame anymore due the lowered buoyancy. *This can clearly be observed from the changing elevation and size of the flame above the tube.* The flame now drops to the inside of the tube where there is abundant H_2 and O_2 available in the restricted tube. The increased pressure and heat cause a huge acceleration in the combustion rate, the internal pressure increases immensely on all sides, the lid gets pushed off and the tube reacts (Newton III) and gets propelled up.

---

2. Produce Floating & Flaming Bubbles

[★★☆☆☆]

This demonstration shows the buoyant nature of hydrogen as well as its exothermic combustion property. It is spectacular in a darkened area. A classroom with standard ceiling heights should work fine.

What you will need

- Balloons filled with **hydrogen**. Connect the balloon to an air cleaning blowgun (see previous chapter)
- Wooden dowel
- Thin wire & pliers
- Cotton fabric piece (old T-Shirt)
- Bubble solution in small cup
- 10 mL (0.33 oz) syringe and plastic tube that fits the nozzle
- Methylated spirits
- Fire lighter
- Ear protection, goggles, lab coat & gloves

Here's how

1. Wrap a small piece of cotton fabric around the end of a wooden dowel. Tie the cloth down with thin wire using pliers.

2. Cut a short length of plastic tubing and fit this to the 10 mL syringe as in Figure 4.9. Connect the other end of the tube to the air blowgun.

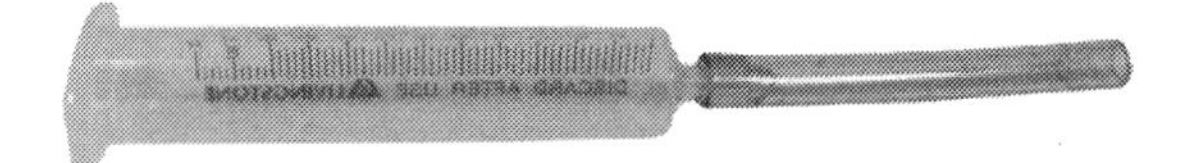

Figure 4.9
Syringe & Tube

3. Connect the H_2 balloon to the blowgun. Dip the <u>open end</u> of the syringe into the bubble solution, remove from the solution and pull the trigger of the blowgun to fill the soap bubble with gas (Figure 4.10). The

bubble will float off after being filled.

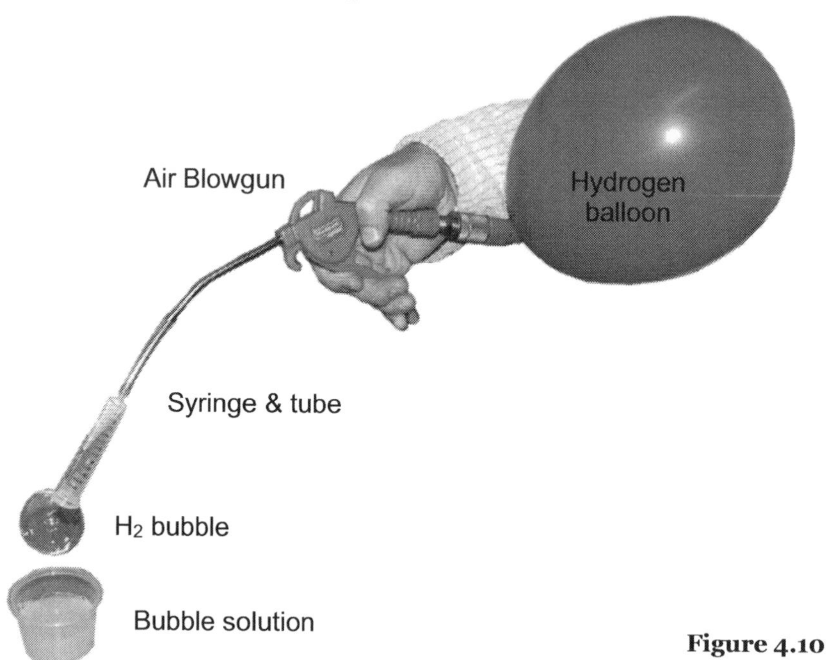

Figure 4.10

4. Prepare a volunteer with safety glasses and lab coat and dip the cotton fabric on the dowel in the methylated spirits. Explain to him/her that he/she should chase the bubbles with the flame and not the balloon! They will require good reflexes.
   Safety: Move the audience back so there is room to operate the flaming dowel and always have a fire extinguisher ready.

5. Dim the lights and ignite the flame.

   Alternatively: Add bubble solution to a plate and produce a number of H_2 domes layered on the plate, using the air blowgun only (no syringe). Move to a safe position and ignite with the flame on the dowel.

What is happening here?

The bubbles are filled with 'pure' hydrogen gas. When a bubble is ignited the flame burns when the gas disperses and mixes with air oxygen. A louder report is obtained by first popping the bubble and then igniting it as more H_2 finds its way to the O_2.

$$2H_2\,(g) + O_2\,(g) \rightarrow 2H_2O\,(g) + \text{energy}$$

3. Produce an Eggsothermic Reaction

[★★★☆☆]

Ready for an *eggs*plosion or an *eggs*othermic reaction?

Back in 1992 Bob Becker described this demonstration using ordinary hen eggs (Ref. 1) but I always had a taste for the bigger spectacle so I used ostrich eggs. Ostrich eggs, due to their large size, produce spectacular eggsothermic demonstrations, but require a polycarbonate safety screen as the egg shell pieces are heavy and projected outwards at high speeds. In the classroom though, ordinary blown-out hen eggs are safer and work extremely well. A safety screen is required too but one that can be easily prepared from a soft drink bottle.

What you will need

- Balloons filled with **hydrogen**. Connect the balloon to an air cleaning blowgun (see previous chapter)
- Large soft drink bottle
- Plastic dome lid with hole (eg. a McDonald® milkshake lid, Figure 4.11)
- Chicken/hen egg(s)
- Retort stand, clamp and boss head
- Hobby knife
- Large paper clip
- Fire lighter (long stemmed)
- Ear protection, goggles, lab coat

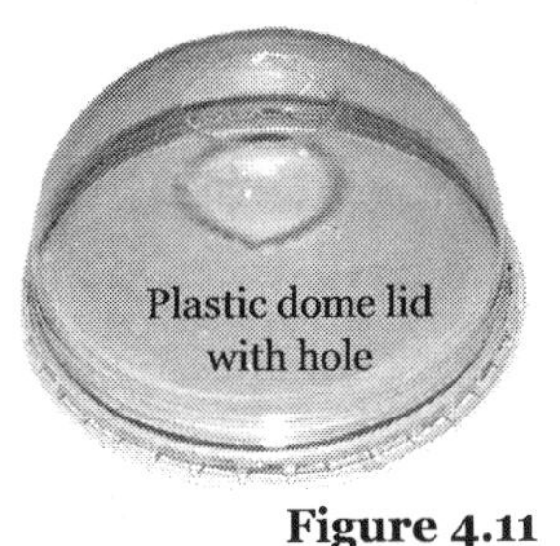

Figure 4.11

Here's how

1. Blow out an egg (or two). How?
   Wash the egg because you have to blow on it. Take a large paper clip, straighten one arm and poke a small hole in each end of the raw egg (the rounded and the pointed ends). Be careful not to crack the shell. Now, poke the clip through one of the holes and pierce the yolk. This makes the egg easier to blow out of the shell. Gently enlarge the hole at the rounded end of the egg to about 1 cm (0.4") diameter. You only need a 1 to 2 mm (1/16") hole on the pointed end. Place your lips over the small hole and steadily blow out the liquid contents into a bowl. Rinse the inner shell under running water and set aside to dry.

2. Prepare a safety screen[1]: Cut a rectangular opening (7 x 9 cm / 3 x 3.5") through the side of a 2 liter soft drink bottle with a knife and position the plastic lid as base on the inside as in Figure 4.12. You may need BluTack® to keep the lid in position. The safety screen is

---

[1] This novel safety screen was demonstrated to me by Jeff Bracken at the Flinn Scientific Morning of Science, NSTA Conference, Philadelphia, March 2010. Brilliant!

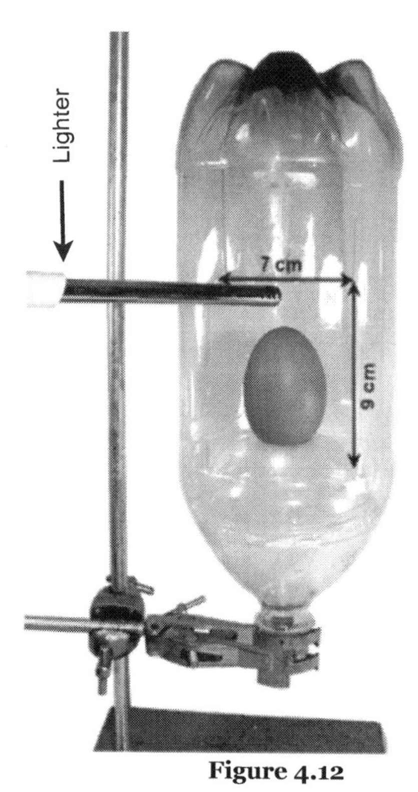

Figure 4.12

clamped to a retort stand using a clamp and boss head. The opening faces the demonstrator . . !

3. Ready for the eggsplosion? *Cover your eyes and ears with protective gear.* Hold the egg in your hand, wet your finger and seal off the top small hole with your finger. Hold the egg upright and insert the blowgun nozzle (connected to the H_2 balloon) into the egg through the bottom hole and squeeze the trigger. The H_2 gas will now displace the air in the egg. Judge the volume of gas delivered by the balloon.

4. Keep the egg upright and sealed with your finger. Hold the lighter in your free hand. Caution the audience to cover their ears. Position the egg on the plastic lid, steady it and gently remove your finger. Now light the H_2 at the top with a long stem lighter so your hand is not close by if the egg explodes. You have a few seconds to light the flame, so take your time . . .

As with the Pringles® Tube you should hear a light 'pop' sound when the gas ignites and if the lights are dimmed, you should see a yellow flame dancing above the hole.

Note: If you did not displace all of the air inside the egg then there is a chance that the egg may explode when you light it. Here's some consolation*: It is a much milder explosion than the Pringles® tube.*

5. Once the flame is lit, step back. There will be a delay from 10 sec to 1 minute and then the egg will pop. A real eggso-thermic change! Although the flame resonates as before it will not be as audible as with the tube.

What is happening here?

The same as with the Tube above.

4. Make an Exploding Boom Box

[★★★★☆]

This is the BIG BOOM - the ideal 'safe' explosion required for a special event. In this demonstration the rapidly expanding pressure wave from the combusting H_2 gas will be encapsulated and amplified by a standard copy paper box. The end result can be quite impressive but is safe as no free flying pieces are encountered.

What you will need

- Balloons filled with hydrogen. You could fill them with lab grade H_2 or prepare your own as indicated in chapter two.
  Note: It is convenient to have these balloons sealed off with balloon clips so you can adjust the size of the balloon
- Cardboard box with lid similar in size to a box that holds 5 reams of copy paper
- Masking tape, packaging tape and string to tie the box
- Matches & sticky stuff (BluTack®)
- General purpose two-core speaker flex wire (at least 5m)
- Thin resistance wire (nichrome wire) ≈ SWG 34; 50 ohm/m (16 ohm/ft)
- Fresh 9V battery
- Ear protection, goggles, lab coat & gloves

Safety / Risk Assessment

★ Be warned. This is a loud explosion if the box is tied up well. Please test the demonstration on your own before performing with an audience.

★ Do NOT add additional oxygen to the hydrogen in the balloon. The oxygen in the box is sufficient to cause combustion. Adding oxygen will increase the output substantially and may cause ear or other physical damage.
Have fun, but please play it safe!

Here's how

1. Prepare the 'igniter'. Clean the insulation from the wires on the two ends of the speaker wire (~ 1 cm). Cut a 8 cm (3") length of nichrome wire and tie it to the wires on one end of the speaker wire (Figure 4.13).
   Test it: Touch the battery momentarily to the other end of the speaker wire to check that it will glow red hot. If not, check the connections.

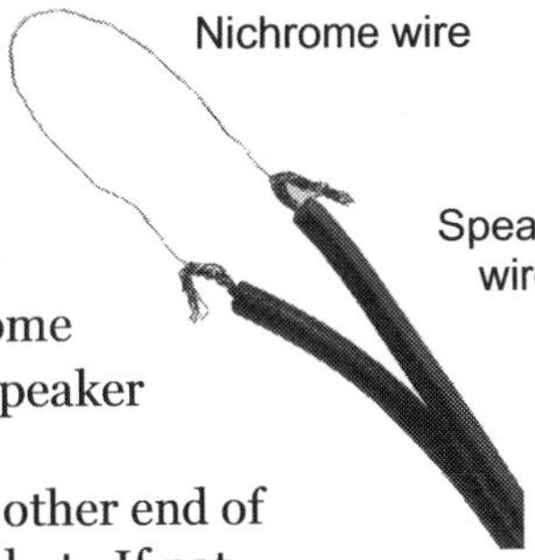

Figure 4.13

2. Bundle the four matches together and tie them up tightly with tape. The match heads should be touching. Now, thread and coil the nichrome wire around the match heads, so it is intertwined with the heads (Figure 4.14).

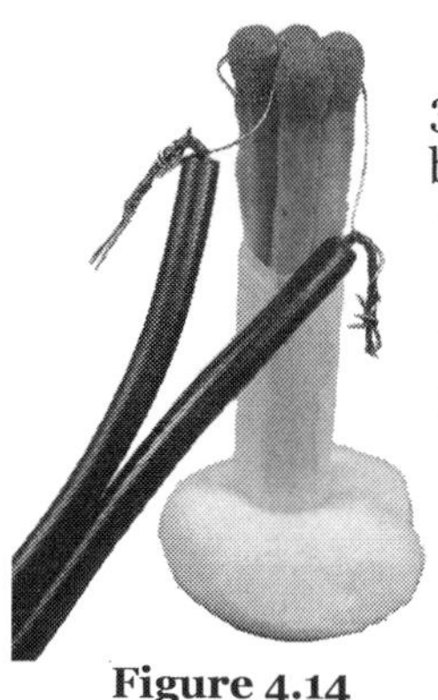

Figure 4.14

3. See figure 4.15. Poke a hole through the corner of the box and thread the speaker wire through the hole, from the inside outwards. The igniter is positioned upright, embedded in a blob of BluTack® close to the corner of the box. Tape the speaker wire down inside the box so it is secure. Check that the nichrome wire is still intertwined with the match heads. Do not test the igniter with the battery as it will be a destructive test!

4. Check that the balloon fits into the box when the lid is closed. If not, release some of the H_2 gas. Then carefully insert the balloon into the box. Check that the enflaming of the matches will happen about 1 to 2 cm below the balloon's surface. If not, adjust the position of the igniter.

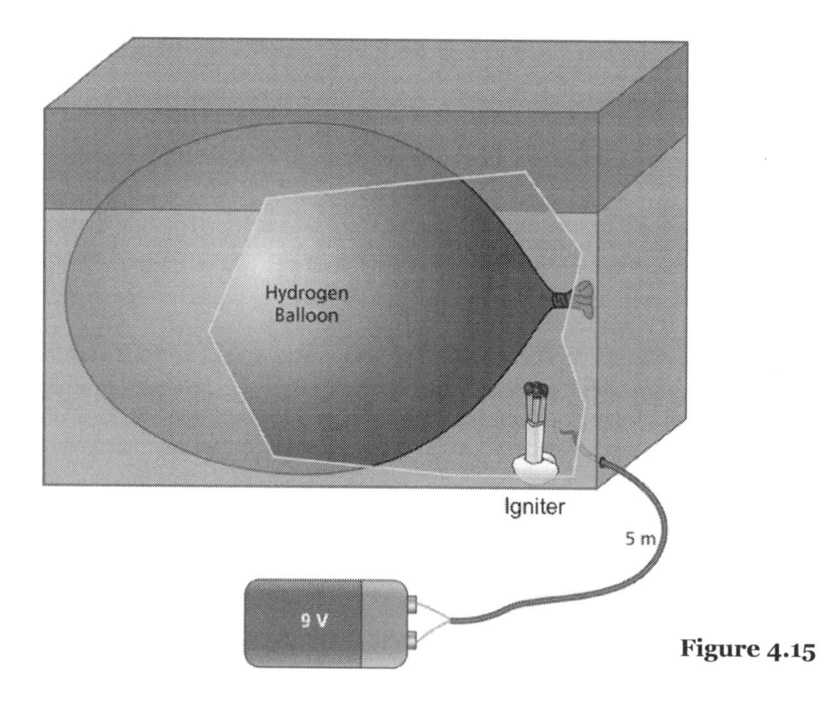

Figure 4.15

5. Close the box lid and **wrap it up tightly** vertically and horizontally with packaging tape (Figure 4.16). You may want to tie it up further with string. The tighter it is packaged the louder the eruption will be. If you did a good job embedding the igniter, then it should still be in position. Done!

6. *Wear ear and eye protection and caution all in the room to protect their ears. All have to stand* ***at least 5 meters off.*** *This includes you!* Have all do a countdown and then touch the speaker wire for 3 seconds to the battery.

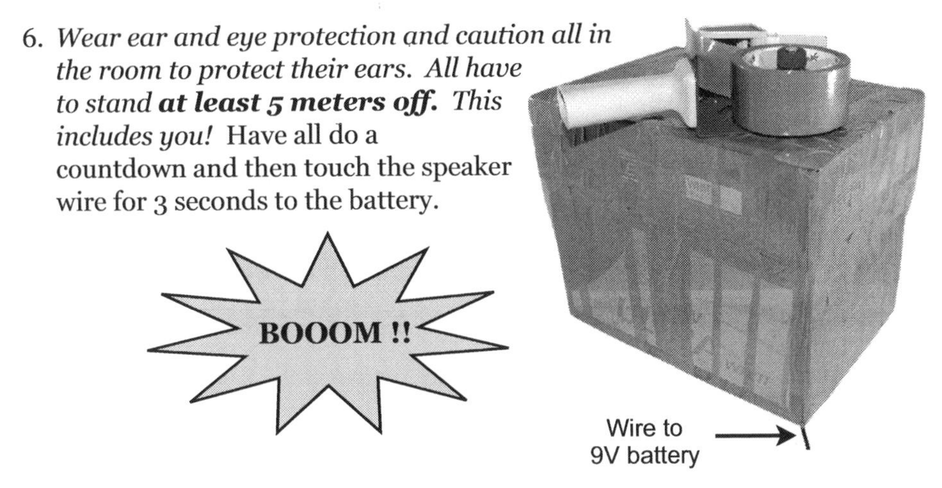

Figure 4.16

What is happening here?

If you ever had the chance to open up a cherry bomb or any other type of firecracker, then you will know that it contains very small quantities of chemicals. The secret of getting the amplified bang from fireworks, lies in the strength of the packaging. When a detonation occurs, the shock wave is amplified hundredfold by simply restricting the rapidly expanding gas volume. The restriction causes secondary detonations, a huge increase in heat and pressure through chain reactions and the end result is the destructive explosion. In the armaments and atomic bomb industries the packaging of the firepower is as important as the chemical composition of the explosives.

Extension

If you have access to firecrackers, wear safety gear and carefully open one. Place the dry powder on a heat pad / tile. Wear safety gear and simply light one end of the pile with a fire lighter. It should cause a small bright flash but no loud sound. Then take a second firecracker outside, light its fuse and experience the bang. Same chemicals, different packaging. Amazing!

Key Terms

Hydrogen chemistry, exothermic reactions, buoyancy, Boyle's Law

References

1. An "Egg-Splosive" Demonstration, Becker, R., J Chem Educ, 1992, 69 (3), p 229 - 230

4 Dihydrogen Monoxide

A most amazing chemical

From a chemical point of view there aren't many chemicals that can beat dihydrogen monoxide, but it is also ranked the most deadliest chemical in the world.

Also known as hydrogen hydroxide, dihydrogen monoxide has caused countless deaths over centuries especially due to accidental inhalation.

In 1989, Eric Lechner, Lars Norpchen and Matthew Kaufman decided to act against this lethal chemical and circulated a Dihydrogen Monoxide contamination warning on the University of California, Santa Cruz Campus. Their campaign was so successful that it was repeated in many other locations globally since then. Have a look at the petition on the next page and try it at your school or neighbourhood and you'll be amazed at the willingness of people to sign!

PETITION

To: All citizens
This petition is for the complete ban of Dihydrogen Monoxide (DHMO).

Dihydrogen monoxide is a colourless, odourless and tasteless chemical. It kills uncounted thousands of people every year. Most of these deaths are caused by accidental inhalation of DHMO, but the dangers of dihydrogen monoxide do not end there. Prolonged exposure to its solid form causes severe tissue damage. Symptoms of DHMO ingestion can include excessive sweating and urination, and possibly a bloated feeling, nausea, vomiting and body electrolyte imbalance. For those who have become dependent, DHMO withdrawal means certain death.

Dihydrogen monoxide:
- is the major component of acid rain
- contributes to the "greenhouse effect"
- can cause severe burns in its gaseous state
- contributes to erosion
- causes corrosion and rusting of many metals
- may cause electrical failures and decreased effectiveness of automobile brakes
- has been found in tumors of terminal cancer patients

Despite the dangers, dihydrogen monoxide is often used:
- as an industrial solvent and coolant.
- in nuclear power plants.
- as a fire retardant.
- in many forms of cruel animal research.
- as an additive in certain "junk-foods" and other food products.

Companies dump waste DHMO into rivers and the ocean, and nothing can be done to stop them because this practice is still legal.

The Government has refused to ban the production, distribution, or use of this damaging chemical due to its "importance to the economic health of this nation." In fact, the navy and other military organizations are conducting experiments with DHMO, and designing multi-billion dollar devices to control and utilize it during warfare situations. Hundreds of military research facilities receive tons of it through a highly sophisticated underground distribution network. Many store large quantities for later use.

Act NOW and sign the petition.

So what is this all about?
Well - simply about H_2O, water or dihydrogen monoxide!

All of the above statements hold true for water. It is just that the selective presentation of the facts are biased and is presented in such a way that you fall into the trap of being persuaded the wrong way. Almost like the political leaflets dropped in our letterboxes so often.

The Background

Water is so common a substance that we often overlook its truly unique properties. For example, given its molar mass, water should be a gas at room temperature. But as we all know, it's not. And it doesn't boil at 25°C (77°F), only at 100°C (212°F) after taking up lots of heat.

Let's have a more unbiased look at some of water's unique properties:

1. Without water **life** as we know it would not exist. It is so vital that NASA launched a multi million dollar probe and plunged it into the moon to determine if the moon has water supplies!

2. Water is never used up - only **recycled**. Your body is almost 60% water by mass and chances are great that some of the billions of water molecules you hold were used by Elvis, Plato or Julius Caesar.

3. Most things shrink when they cool, but not with water. Water **expands** as it starts to freeze. When you cool down water, the molecules move closer together but at a temperature of 4°C (39°F), the molecules are as close as they can get (maximum density). With further cooling they start to rearrange themselves into a more open structure, that of ice (Figure 5.1). Ice is therefore less dense than liquid water and floats on water (0.9167 g/mL at 0°C). This is an expansion of nearly one tenth its original volume.

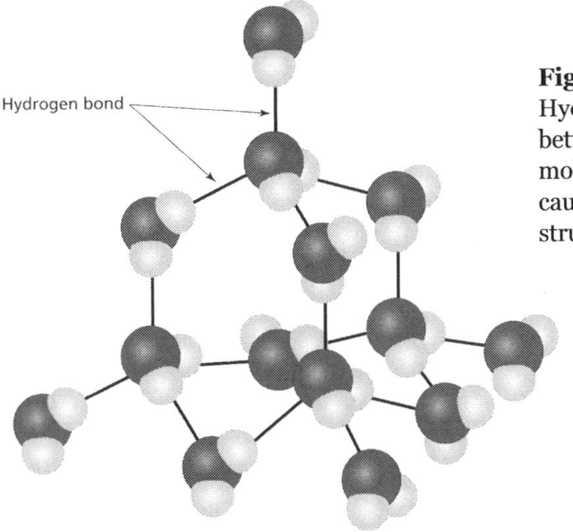

Figure 5.1
Hydrogen bonding between water molecules in ice causes an open structure

4. There is an invisible film on water's surface that supports insects and make water molecules group together in droplets. We call it **surface tension**.

5. Water can climb up a tube against gravity using its adhesion capabilities in a process known as **capillary action**.

6. Water has a high **specific heat capacity** (C_p) and that means it can hold or store more heat per mole of substance than any other compound (with the exception of ammonia). The high heat capacity is mirrored in its high boiling point compared to other hydrides of Group 16 in the periodic table. In practical terms this means that water can act as a storage facility of heat without raising its temperature substantially. In simple terms: Water has a huge affinity for heat!

   Looking at Table 5.1, hydrogen has a much higher C_p value than that of water. But this does not paint the full picture - when converted back to a molar (per molecule) basis, $C_{p,m}$, water is still the best heat absorber.

Substance	Specific Heat Capacity, C_p $Jg^{-1}K^{-1}$	Specific Heat Capacity $C_{p,m}$ $Jmole^{-1}K^{-1}$
Copper	0.39	24.47
Iron	0.45	25.10
Hydrogen	14.3	28.82
Air	1.012	29.19
Water	4.18	75.33

Table 5.1

It is this high heat capacity of water that is often taken for granted that I would like to briefly explore in this chapter. Why does it demonstrate this property? Why is this property so important? What are the practical outcomes?

Heat Capacity

From the C_p values in Table 5.1 it follows that 1 kg of water can absorb ("drink") almost 9.3 times and 10.7 times as much energy as iron and copper, *before it raises its temperature with 1 degree.* It is interesting that most metals may be better heat conductors than water but none come close when it comes to storing heat in their heat reservoirs.

So why can water do this?

Water is that unique a chemical because of the **geometry in which water's three atoms are bonded!** The three atoms are arranged in a triangular shape. The triangular arrangement causes an imbalance in the way in which bonding electrons are shared, causing the oxygen end of each water molecule to carry a slightly negative charge and the hydrogen ends to be slightly positively charged. We therefore call water molecules **polar**.

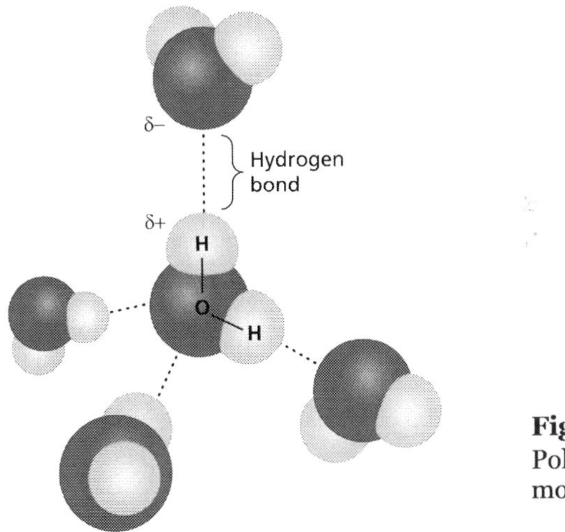

Figure 5.2 Polar water molecules

The positive, hydrogen end of one water molecule will attract to the negative, oxygen end of other water molecules (Figure 5.2). This is what makes water molecules stick together through thick and thin. We call it **hydrogen bonding**, a relatively weak **intermolecular force** acting only over very short distances. Although it is relatively weak (4 - 40 kJ/mole) when compared to ion-ion interactions (eg. Na^+Cl^-, 40 - 400 kJ/mole) or covalent bonding (200 - 800 kJ/mole), it explains all of water's unique properties. Because of these bonds, lots of energy is required to break water molecules apart. Its capillary action, surface tension, high boiling point, heat capacity and heat of evaporation is primarily due to hydrogen bonding.

But there is another reason why water behaves as it does.
Specific heat capacity, C_p, is the capacity for heat ("storage ability") that a certain quantity of a substance has. It is a function of the **structure of the substance** but depends in particular also on the **number of degrees of freedom** that is available to the particles in the substance when absorbing heat.

When molecules absorb heat, they undergo many characteristic internal *translational* and *vibrational changes*. Their energy content is stored two-fold: as kinetic energy (manifests as translational motion and a temperature change) and as thermal potential energy (the degrees of vibrational freedom available to particles). More internal degrees of freedom for particles tend to increase a substance's ability to store heat and thus its specific heat capacity.

As many Infra Red spectroscopists know, a molecule can transform its absorbed energy into any of many "degrees of freedom". There is one degree of freedom per independent mode of motion, ie., translational (whole body movement in the x, y or z directions), rotational (whole body rotation around one or two independent axes) or vibrational (oscillation of bond length or angle).

	Translational	Rotational	Bond length	Bond angle
O_2	3	2	1	
H_2O	3	3	2	1

Table 5.2

For water, each molecule has 9 degrees of freedom which accounts for its increased specific heat capacity (Table 5.2).

One of these degrees of freedom has a very practical outcome in our everyday lives. The intrinsic energy of **microwaves** correspond to that of water's molecular rotational modes. A microwave simply excites the rotational mode of water and fat molecules. The absorbed energy then dissipates as heat (translational motion), and the food's temperature increases. This is why when water boils in a microwave oven the cup remains relatively cool.

So water has an additional capacity for heat storage in the form of *vibrational potential energy*, even at low temperatures (ie. near the freezing point of water).

All of these micro-scale characteristics confirm the practical experience we have with water on a macro-scale. It simply is a great heat storage medium (coolant) and heat sink. Here are a few of its heat absorbance applications:

- It is estimated that water in the earth's oceans absorb one thousand times more heat than the air in the atmosphere and is holding 80 to

90% of global warming heat without raising its temperature [1]

- It is the chemical of choice in most heat exchange processes in the world. In most industrial countries the bulk of industrial water usage is for cooling
- Our car engines have water in their radiators and they won't last much longer than 20 minutes without water's magical absorption
- Immediate application of cold water to any burn wound (except for sulphuric acid burns) will see heat absorbed from the wound by the water
- When bush fires approach people soak their property in water so it can act as a heat 'sink' for the approaching heat
- Sweating is an effective way to regulate body temperature. This is because of the high molar heat of vaporization (41 kJ/mole) of water. You require more than five times the heat required to heat a quantity of water from 0°C to 100°C (32 to 212°F) to vaporize it. On average, a 60 kg person generates 1×10^7 J of heat per day from metabolism. If sweating were the only way to maintain body temperature, then 4.4 litres of sweat would be required. Fortunately body heat is also radiated to the cooler environment.

Let's now look at the fun practical part.

A. The Balloon-and-Water Trick

[★☆☆☆☆]

This bar-trick has been around for some time but it never stops amazing people.

Safety

★ Do not use open flames in the proximity of flammable gases or combustible material

★ Have a fire extinguisher readily available

What you will need

- Rubber balloon
- Fire lighter (long stem) or candle
- Water from a tap (faucet) and a sink

---

[1] http://www.nasa.gov/multimedia/podcasting/jpl-earth-20090421.html; Dec 2010

Here's how

1. Pull the balloon's mouth over the tap (faucet) and fill the balloon with water (about ¾ cup (6 fl oz)).

2. Remove the balloon and adjust the volume so that you have about 120 mL (half a cup or 4 fl oz) by squirting water from it. Tie it and put the balloon aside for later use.

3. Blow up another balloon without water in it.

4. Ask students what will happen when you heat rubber and air and then heat the balloon without water. POP!

5. Now pick up the balloon with water. Hold the balloon upright and (if you dare to) above the heads of members of the audience. Light the lighter and hold the flame under the part of the balloon that is in contact with water (Figure 5.3). Amazingly it does not pop.
   If you use a candle check that you do not touch the balloon with the hot wick. The heat is too concentrated and will melt a small hole in the balloon. Try it sometime!

Figure 5.3

6. Feel free to hold the flame there for a couple of seconds while explaining the wonders of water to the audience. I have held it for up to 3 minutes before it popped. But this will depend greatly on the balloon quality and amount of water.

Discussion

In the first balloon we had air that has a low heat capacity ($C_p = 1.0$), that couldn't absorb heat fast enough so the heat was used to melt the rubber. In the second balloon the water ($C_p = 4.18$) absorbed the heat at such a rate that no heat could be used for melting the balloon. Note that even hot

water will absorb heat amazingly well (as in the car engine). The C_p of water at 100°C (212°F) is 4.22 J/gK. Due to its high **heat capacity** the water's temperature did not increase substantially.
There is another factor, though, that plays an important role here; that of the **heat transfer coefficient** which is much higher for a liquid than for a gas. This factor is incorporated by engineers when designing and evaluating heat exchanger efficiencies. The heat transfer coefficient for **air** is 10 to 100 W/m^2K and that of **water** 500 to 10,000 W/m^2K.

B. Boiling water in a Paper Pan

[★★☆☆☆]

This is one of the simplest but most impressive demonstrations one can do. It surely delivers a great discrepant event.

Safety

- ★ Do not use open flames in the proximity of flammable gases
- ★ Have a wet cloth and a heat-resistant surface available to snuff the fire when the paper pan combusts
- ★ Have a fire extinguisher readily available
- ★ Wear gloves and glasses
- ★ Take care with the hot water

What you will need

- Two sheets of standard copy paper (usually 80 gsm)
  (You may use a small paper cup and paper bags [2] but I think the power of this demonstration lies in the use of 'ordinary' copy paper to boil water!)
- Tripod with metal gauze
- Bunsen burner or gas camping stove
- Office stapler
- Heat resistant surface
- Wet towel
- Water in trough

Here's how

1. Draw two squares on both sheets of paper as in Figure 5.4 with sides of 13 cm (5.2") and 18 cm (7").

2. Cut the outer square from both papers.

---

[2] It is important that the flame, water and paper surface be in close contact so heat transfer can take place. This can be a problem when using thick cardboard.

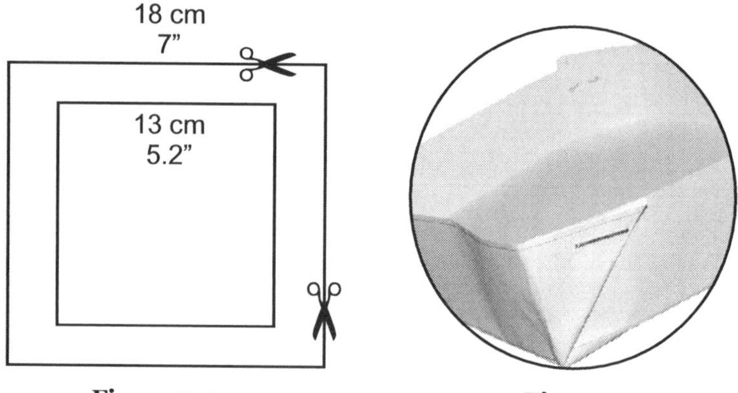

Figure 5.4 **Figure 5.5**

3. Fold the paper back on the 13 cm (5.2") lines and fold the corner flap sideways so you get a right angled corner (Figure 5.5). Staple the flap to a side wall. Do this at all four corners.
   Ask your audience if they think it is possible to boil water in a paper pan while showing the flimsy pan to them.

4. Light the gas, heat the metal gauze to red hot and place the empty 'paper pan' on the gauze (Figure 5.6). Have a wet towel and water ready as it <u>will</u> catch fire. Snuff the flames with water and express your disappointment with the outcome of the experiment.

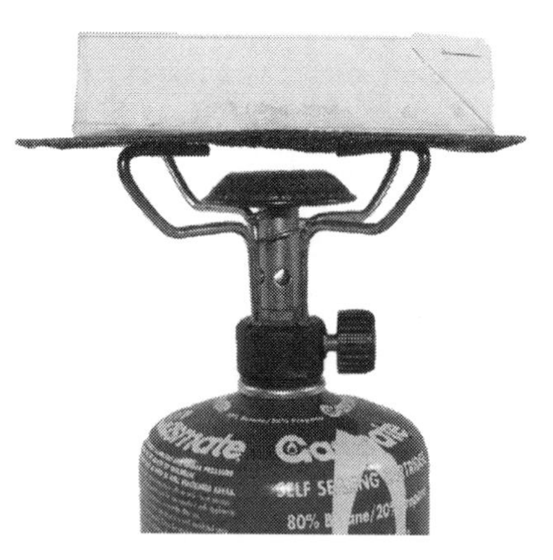

Figure 5.6

5. Now stress that you may have used the wrong procedure. Again, heat the gauze to red hot. <u>First</u> fill the second pan ⅓ full with water and then lift from the two ends and place on the red gauze . . . no paper combustion!

   Now is *the time to discuss water's enormous heat capacity. You may want to use the analogy with a slow flowing tap leaking water into a sponge. The sponge will only release the water once its "water capacity" is exceeded. Same with water & heat. There is no combustion of the paper as long as the heat reservoir still has absorbance capacity.*

Figure 5.7

6. Draw students' attention to the steam forming and when the water starts boiling, pour yourself a cup of coffee and drink to water's "invisible safety features". Focus attention on the unscathed bottom of the pan (Figure 5.7)!
   I call this *cool hot* science!

Teaching Extension

- Place a raw egg in the water in the paper pan and boil for 4 minutes or how about a poached egg? (Figure 5.8) Another eggsothermic process - enjoy a science breakfast!
- Alternatively you may use a thin paper cup or paper bag. If the bag leaks, seal it with tape.

Figure 5.8

Discussion

Although this demonstration points to the same phenomena of heat capacity, it is different in that the water is heated to boiling point. So we have an additional "sink" for the flame's energy - that of *latent energy* during a *phase change*. When the water phase change starts (liquid → gas) the heat capacity is infinite!

Let's sum up:

When we boil water it absorbs heat

- ✦ to break its intermolecular hydrogen bonds and cause molecular vibrations that dissipate as heat (translational motion) raising its temperature to boiling point and
- ✦ to facilitate the escape of molecules to the gas phase (phase change → latent heat of evaporation).

Both these 'heat sinks' ensure that the paper's temperature stays below its ignition temperature of 233°C (451°F). This is a remarkable outcome as the temperature of the flame is anywhere between 320°C and 1600°C (*Ref. 4*)

C. Igniting Bubbles on your Hand (safely) [★★★★☆]

Many teachers are familiar with the 'non-burning' American dollar bill or handkerchief demonstration (*Ref. 2*). In this demonstration a US dollar bill is set alight but the paper is not consumed as the fuel is mixed with water. This is another great demo on water's insatiable appetite for heat!

In the next demonstration we will rely heavily on water's magical property of heat absorption and find that it will protect us from the flame's heat. We will ignite a small quantity of butane bubbles on a gloved hand! It is much safer than it appears and clearly demonstrates the amazing heat capacity water has. Need I say that it is very popular with students of all ages - but do warn them that it is a specialized demonstration and not to be practiced on their own.

Figure 5.9
Water will absorb heat from flames and thus protect. When bush fires approach, people sometimes spray buildings with water and cover themselves in wet blankets.

Safety / Risk Assessment

- ★ Do not use open flames in the proximity of flammable gases
- ★ Check that you are at a safe distance from flammable material such as paper, curtains, etc.
- ★ Ensure that you wet the gloved hand and exposed body parts thoroughly with water. Follow instructions closely
- ★ Have a fire extinguisher readily available
- ★ Wear safety glasses and a lab coat
- ★ Only use a small volume of bubbles

What you will need

- A butane gas refill cartridge sold at tobacconists or LPG (butane/ propane) gas cartridge (Figure 5.10).
  You may use methane, propane or butane gas. Their BTUs are in a similar range[3]. Natural gas (methane gas) is found in many laboratories. It is less dense than air so may produce an interesting 'towering effect'
- Brass tubing, same diameter as gas cartridge metal tube, 3 cm (1.2") length (Figure 5.11)
- Plastic tubing to fit over brass tubing, about 1 m (3.3')
- Latex glove
- Bubble solution
- Water in open container

Here's how

1. I prefer to use a mobile gas supply such as a butane gas refill cartridge. The gas is only released by the cartridge when pressure is applied. Insert the brass tube into the plastic tube (Figure 5.11). This will trigger the pressure release valve when you exert pressure on the valve and the gas will enter the tubing.

Figure 5.10 Figure 5.11

2. Place bubble solution in an clean open container so the gas tube can easily enter it. It should be on a smooth, clean work surface.

3. Open the gas flow by squeezing the brass tubing onto the gas cartridge outlet tube. Bubbles will start frothing and building in the container and overflow the container (Figure 5.13).

---

[3] The British thermal unit (BTU) is a traditional unit of energy equal to about 1.055 kJ

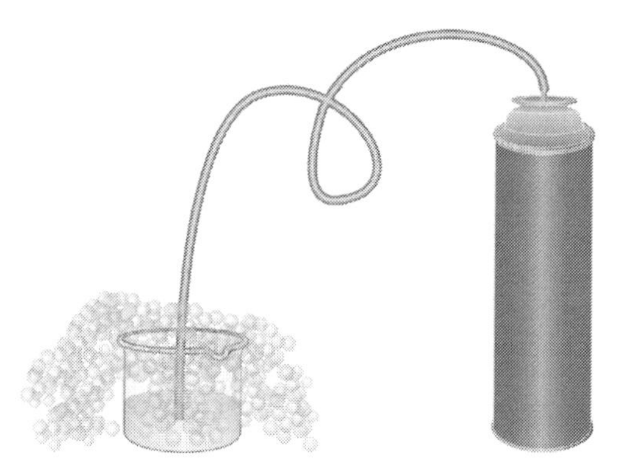

Figure 5.12

4. Once you have collected enough bubbles, close the gas supply and remove the tube and cartridge.

5. Fit the glove on one hand and then dip both hands in water. Splash any exposed parts of your arms with water too.

6. Scoop **a small quantity of bubbles** onto your gloved hand and move to a safe position away from paper and other flammable stuff. **Extend your hand away from your body.**

Figure 5.13

7. Light a fire lighter with your free hand and bring the flame slowly closer to the bubbles . . . WHOOSH! The flames should extinguish automatically once all gas is consumed.

Key Terms

Specific heat capacity, latent heat, heat of evaporation, exothermic reactions, combustion, hydrogen bonding

References

1. Chang, R., Physical Chemistry for the Chemical & Biological Sciences, University Science Books: Sausalito, 2000
2. Shakhashiri, B. Z.; Chemical Demonstrations: A handbook for teachers of chemistry, Vol. 1, The University of Wisconsin Press: Madison, 1983; p 13
3. Koehler, K. R. (1996), Infrared Spectroscopy. http://www.rwc.uc.edu/koehler/biophys/6e.html; Dec 2010
4. Gaydon, A. G., Wolfhard, H. G., Flames: Their Structure, Radiation and Temperature, 3rd ed., Chapman and Hall, London (1970).

5 Silicon from Sand

Next time you step onto the beach, bend down, grab a handful of sand and admire the fact: By mass 47% of what you hold in your hand is the element silicon. The rest is simply oxygen. Remarkable!

Silicon is the second most abundant element in the earth's crust (27.7%) - only oxygen beats it - and can easily be extracted from white sand (SiO_2) in a spectacular reaction in the school science laboratory.

The Background

Silicon

Of all the elements, silicon is the closest chemically related to its neighbour carbon (the element of life) and can therefore form similar tetravalent covalent bonds. It is however much less reactive than carbon. Being a *metalloid*, its properties are intermediate between that of metals and non-metals.

Silicon does not occur free in nature. It mainly occurs in minerals consisting of pure silicon dioxide (SiO_2, silica) in different crystalline forms (quartz, mica, opal) and as silicates. These minerals are found in clay, sand and various types of rock. Sand is basically tiny crystals of silica.

Silicon is the principal component of most semiconductor devices (eg. solar panels), glass, cement and ceramics. It is often confused with the polymer substances known as *silicones*, which do contain silicon but are soft and rubber-like. Silicones have found uses as medical implants, bathroom sealants, silly putties, etc. When silicon is combined with carbon, it forms silicon carbide (SiC) or carborundum, the next hardest substance after diamond. Furthermore, quartz crystals keep the time in our modern watches as its oscillations are extremely stable.

Silicon is widely used in semiconductor processors as it forms n and p type transistors (MOSFETs) when bombarded with boron and phosphorous ions. The high-tech Silicon Valley region in California is aptly named after this element. Surely there would be no Microsoft®, Apple® or wealthy IT magnates without silicon. This once again confirms that Chemistry rules!

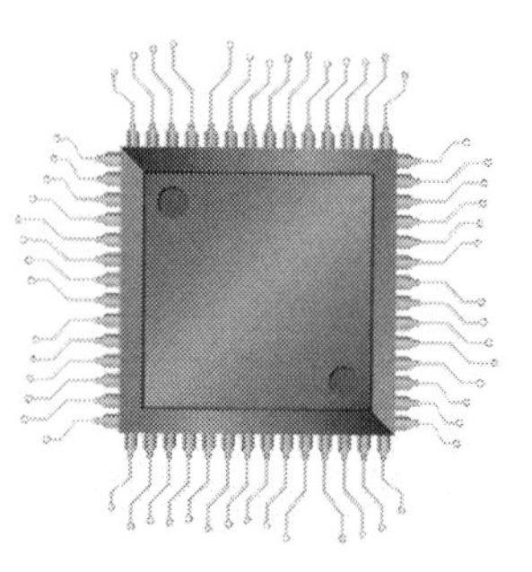

Characteristics of Silicon

In its crystalline form, silicon has a dark gray colour and a metallic lustre. It forms four covalent bonds with four other silicon atoms very similar to the hard diamond structure of carbon. It is not as hard due to the larger size of the Si atom which causes increased inter-atomic distances and thus weaker bonds. Even though it is a relatively inert element, silicon still reacts with halogens and dilute alkalis, but it is not affected by most acids (except conc. HCl).

Production of silicon

In order for us to get to pure silicon, the silicon has to be reduced from silica - the oxygen has to be removed. This is accomplished by heating a mixture of silica and carbon in the form of coke in an electric arc furnace using carbon electrodes. At temperatures over 2,000°C (3,600°F), the carbon reduces the silica to silicon:

$$SiO_2 + C \rightarrow Si + CO_2$$

Liquid silicon settles in the bottom of the furnace and is then drained and cooled. The silicon produced via this process is called *metallurgical grade silicon* and is up to 99% pure. This is not nearly pure enough for semiconductor manufacture so it is further refined with gaseous hydrochloric acid, fractionally distilled and reacted with hydrogen at 1,100°C. A complex crystal transformation process follows and the silicon rod known as a 'boule' is sliced up into wafers typically 0.775 mm thick.

The electronic properties of the wafer is then modified through exposure to ion beams, UV light, hot gases and chemicals. A high-tech, complex "Silicon Valley" type of activity.

The Reduction of Metals from their Oxides

Thermite reactions

In Thermite reactions **metal oxides** react with **aluminium** to produce the molten metal. It requires substantial activation energy to get going.

These highly exothermic redox reactions are known as the Goldschmidt reactions (after Hans Goldschmidt who developed them in 1893).

They have been used industrially for welding (even under water), the preparation of metals from their oxides (reduction) and the production of incendiary devices. The process is initiated by heat but then becomes self-sustaining.

Figure 6.1

In the early 1900's a product called Thermit® was developed and used worldwide to weld rail tracks. The photo[1] in Figure 6.1 was taken in 2010 proving that it is still in use today. Here is the reaction:

$$Fe_2O_3 + 2Al \rightarrow 2Fe + Al_2O_3 + \text{energy}$$

The question is: Can silicon be persuaded to give up its oxygen in a similar thermite process?
For Fe_2O_3, ΔH_f = -822 kJ/mole; for SiO_2, ΔH_f = - 859 kJ/mole

In 1902 Kuhn (Ref. 2) described a method in which a primary reaction provides the activation energy to initiate the secondary reduction of silicon from silica.

Extracting Silicon from Sand

[★★★★☆]

Crystalline silicon can be prepared using a *variation* of the thermite reaction in which sufficient activation energy is produced between aluminium and sulfur.

---

[1] Photo by Joliet Jake Szerkeszto from WikiMedia Commons

Safety / Risk Assessment

★ Burning magnesium produces **bright light** that may cause temporary loss of sight. Avoid looking directly at the flare
★ Magnesium is very **reactive** and contact with other chemicals may result in explosion
★ This demonstration produces **intense heat** and molten silicon. A dry-powder fire extinguisher should be readily available at all times. DO NOT use **water** as an extinguisher as this will produce potentially explosive hydrogen gas
★ The near-impossibility of smothering and the high temperatures generated make thermite reactions potentially hazardous. Appropriate precautions must be taken **before** thermite is ignited. Keep all flammable material away from the area
★ The reaction should be performed in a fume cupboard or outdoors behind a **safety shield** as it produces intense heat, smoke and molten metal. Sparks can fly up to 2 m horizontally
★ Wear heat-protective welding gloves and use metal tongs to handle the fragments of the clay pot and molten silicon after the reaction has taken place
★ Wear protective clothing and safety glasses

Chemical Safety

Aluminium (Al)
Risk phrases: R10 - 15 Extremely flammable; Contact with water liberates highly flammable gases.
Safety phrase: S7/8/44 Keep container tightly closed; Keep container dry.

Magnesium (Mg)
Risk phrases: R11-15 Extremely flammable; Contact with water liberates highly flammable gases.
Safety phrase: S7/8-43 Keep container tightly closed; Keep container dry.

Hydrochloric Acid (HCl) is a strong acid. Treat with the greatest respect.
Risk phrases: R34-37 Causes severe burns; Irritating to eyes & respiratory system.
Safety phrase: S2-26 Keep out of reach of children; In case of contact with eyes, rinse immediately with plenty of water and seek medical advice; Wear suitable gloves and eye/face protection.

Sodium hydroxide (NaOH) destroys clothes and causes injury to the skin. Treat with the greatest respect.
Risk phrases: R35 Causes severe burns
Safety phrase: S2/26/37/39 Keep out of reach of children; In case of contact with eyes, rinse immediately with plenty of water and seek medical advice; Wear suitable gloves and eye/face protection.

What you will need

Chemicals

Figure 6.2

- Fine aluminium powder, 325 mesh or finer
- Powdered sulfur
- Dry white sand - beach sand or washed white builder's sand
- Hydrochloric acid, 4M (add 180 mL conc. HCl to 500 mL water)
  Initiator chemicals:
- Fine magnesium powder & magnesium ribbon (5 cm) ***or***
  Potassium permanganate ($KMnO_4$) and glycerine
- **Extension chemicals:**
  Sodium hydroxide solution, 5M (dissolve 50 g NaOH in 250 mL water)

Other

- Mortar & pestle
- Small clay flower pot, ~ 6.5 cm (2.5") inside top diameter
- 100 mL glass beaker & plastic container
- Electronic balance, 100 g ± 0.1 g
- Heat resistant pad (1 m x 1m)
- Transparent safety shield or fume cupboard
- Heat-protective gloves eg. welding gloves
- Spatula & metal tongs
- Fire lighter with long stem

For Extension activities

- Test tube, clamp & burner
- Wires with croc clips (2)
- Ammeter & battery pack
- 3.2 V lamp in holder
- Hammer & Nail

Here's how

1. Dry about 50 g of sand in an oven for 2 hours (~ 180°C, 350°F) and then grind the sand to a powdered form in a mortar and pestle.

2. Weigh and mix the following to form a homogenous mix:

   8 g of dry aluminium powder
   10 g of powdered sulfur
   7 g of dry powdered sand

Figure 6.3 **Figure 6.4**

3. The best mix is obtained by shaking the mixture for a minute in a sealed plastic container (Figure 6.4). Do NOT grind them together in a mortar and pestle.

Figure 6.5

4. Use the small clay flower pot. If it has a hole in the bottom - cover the hole with paper. Transfer all of the mix to the pot (Figure 6.5).

5. Make a small cone-shaped indentation at the top (2 cm deep and 2 cm wide) and fill this with magnesium powder to facilitate ignition. (For an alternative method see Note 1 below). Fray a 5 cm (2”) magnesium ribbon’s end with scissors. This will enlarge its surface area and aid ignition. Push this fuse into the magnesium powder pile.

Safety: Position the pot on a heat resistant pad in a fume cupboard or outdoors away from any combustible material. Use a safety shield. Wear protective eye wear, gloves and a lab coat.

6. Light the frayed end of the magnesium ribbon using a long stemmed fire lighter (Figure 6.6). It may take a few seconds before the ribbon starts burning. Once this happens – step back immediately. Ignition of

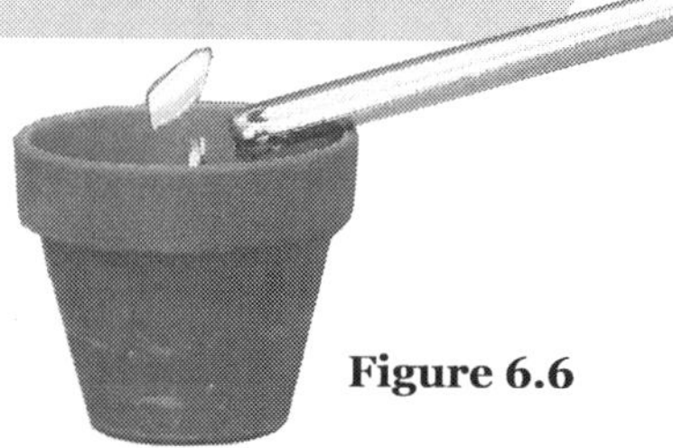

Figure 6.6

the mixture is not instantaneous and might take as long as 60 seconds. The reaction will produce lots of **spattering, bright light** and **intense heat.** The temperature is reported to be well above 2,200°C. The residue will have an orange-red glow and be hot for some time. If required, carefully pick up the red-hot residue using metal tongs.

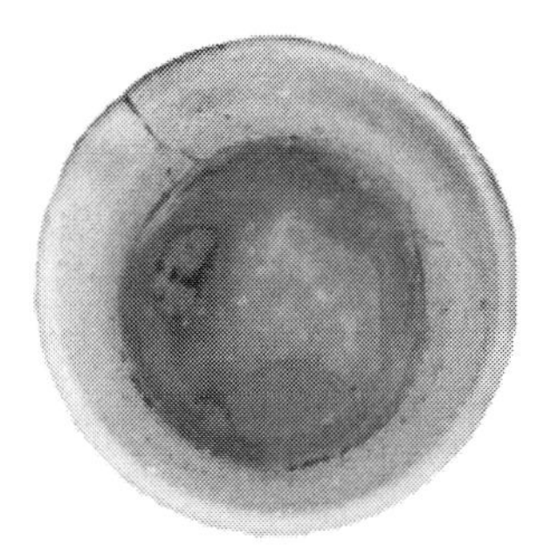

Figure 6.7
The clay pot as seen from the side and top

If the reaction fails to react: Wait 3 minutes. Do not approach the reaction vessel until you are sure ignition is not possible. Replace the ribbon or add more magnesium powder. Try again.

Note 1: You may prefer to use potassium permanganate ($KMnO_4$) and glycerine as initiators in stead of magnesium. In this case a small quantity (~ 1 to 2 teaspoons) of the finely powdered permanganate is inserted in the cone-shaped indentation in the sand mix. Form a small depression in the top of the permanganate. When you are ready to start, a few droplets of glycerine is added to the depression with a dropper. Step back and wait. *The speed of this reaction is closely related to the permanganate solid size.*

7. Purify the silicon residue.
   Leave the clay pot to cool for 20 minutes. Break the pot apart and separate the silicon residue from the clay pieces (Figure 6.8). Break the residue into smaller pieces using light taps from a hammer.

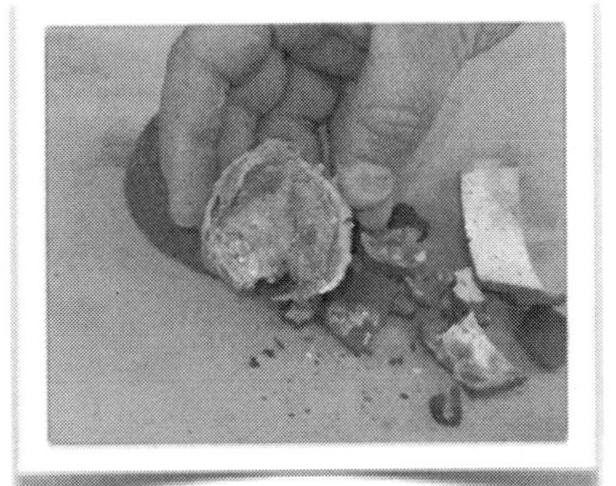

Figure 6.8

The Chemistry

There are basically three reactions here that produce a chain reaction effect.
Magnesium burns in oxygen and produces the activation energy to get the sulfur and aluminium going.

$$2Mg + O_2 \rightarrow 2MgO + \text{energy}$$

The aluminium and sulfur again, react in an exothermic reaction that is the source of activation energy for the silica and aluminium.

$$2Al + 3S \rightarrow Al_2S_3 + \text{energy}$$

$$3SiO_2 + 4Al \rightarrow 2Al_2O_3 + 3Si + \text{energy}$$

8. The silicon (Si) and aluminium oxide (Al_2O_3) is now separated from the aluminium sulphide (Al_2S_3) by adding the residue to diluted hydrochloric acid in a glass beaker (acid - take care!). Copious amounts of **hydrogen sulphide** will be evolved – foul smelling rotten egg gas - use a fume cupboard or perform outside:

$$Al_2S_3 + 6HCl \rightarrow 2AlCl_3 + 3H_2S$$

   Leave the residue in the acid until gas formation subsides. This process may take up to 20 minutes. The finer the residue, the faster this reaction.

9. Now, carefully discard the acid under running water and wash the residue in the beaker for 30 seconds. The silicon will be found in the form of pea-size, hard, black, **crystalline globules**. It is insoluble in most acids. Boil the globules in a glass beaker using dilute hydrochloric acid for further purification.

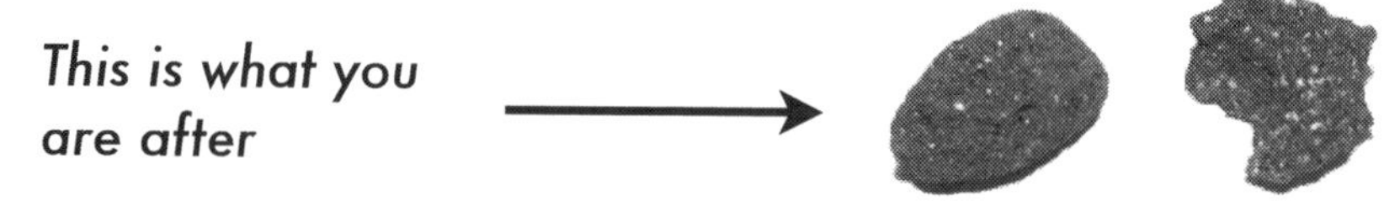

Figure 6.9

Yield calculations

Students can weigh the silicon and determine the yield:
This is a typical result for this reduction:

$$\text{Yield} = 1.1\text{ g silicon} / 7\text{ g sand} \times 100$$
$$= 16\ \%$$

Teaching Extension

How do we know the globules are silicon?

Any of the following methods can provide evidence that the globules are indeed silicon:

A. Silicon's reaction with caustic soda

Silicon is insoluble in common acids but will react with alkalis liberating hydrogen gas. This provides a test for silicon. Place a cleaned small silicon globule in a test tube. Add 5M sodium hydroxide solution (caustic soda) and heat gently for 1 – 2 minutes. The characteristic 'pop' when a lighted splint is brought to the mouth of the test tube will confirm the presence of hydrogen.

$$Si + 2NaOH + H_2O \rightarrow Na_2SiO_3 + 2H_2$$

$$2H_2 + O_2 \rightarrow 2H_2O + \text{energy}$$

B. Demonstrating the reverse temperature effect

Clamp one cleaned globule of silicon using two crocodile clip wires without plastic insulation. Check that the clips hold on firmly but do not touch each other (Figure 6.10).

Figure 6.10

Connect the following items in series to the two wires:

- ★ One battery pack: 2 x AA (3.0V)
- ★ One 3.2 V lamp in a lamp holder
- ★ One ammeter (multimeter)

The lamp should light up. Record the ammeter reading.

Now **heat the silicon** in a bunsen flame or blow torch.
The lamp should brighten up and the **ammeter** reading should **increase** (increased current).
Silicon is a semi-conductor. It has a reverse *temperature-resistance coefficient (resistance decreases as the temperature increases) since the number of free charge carriers increases with temperature.*

To compare with a conductor, find a metal such as an iron nail and repeat the process. Metals (conductors) experience an increase in resistance with increasing temperature and thus a decrease in current.

Disposal

Allow the solids and clay fragments to cool to room temperature. The clay flower pot invariably cracks and should not be re-used. Dispose of all solids in the waste bin.

Key Terms

Silicon, metal reduction, thermite reaction, metalloid, semi-conductor, activation energy, reverse temperature effect.

References

1. *Bedford, M (2009) How sand is transformed into silicon chips. www.techradar.com, Dec 2010*
2. *Grant, R. (June 1979) Some experiments with silicon, Spectrum, 17, pp 19 - 20*
3. *Shakhashiri, B. Z.; Chemical Demonstrations: A handbook for teachers of chemistry, Vol. 1, The University of Wisconsin Press: Madison, 1983; pp 85 - 89*
4. *"Values of Chemical Thermodynamical Properties", CRC Handbook of Chemistry and Physics, 57th Edition. CRC Press*

6 Magnesium Pencil Sharpener Secrets

This is the stuff MacGyver's dreams were made of.
If your students think the sodium and potassium-on-water demonstration is spectacular, wait till they see this one. Magnesium beats the other alkali metals hands down in this eye popping spectacle.

Can iron burn? Can you light steel with a battery? Can magnesium burn under water?
These questions are not only intriguing, they touch the basics of alkali and alkali-earth chemistry and make great visual demonstrations in the laboratory.

We have mentioned before that combustion (burning) is simply a chemical reaction in which a chemical substance reacts rapidly with oxygen to form an oxide, heat and light in an exothermic reaction. It is also known as 'oxidation' since an oxide is formed and classified as a redox reaction. Thankfully it requires activation energy to get the process going eg. lightning, flame, spark or friction.

Lets look at the first question.

A. Can Iron burn?

[★★☆☆☆]

What you will need

- Steel wool - In paint section at hardware store, Grade: 0000, super fine
- Steel nail
- Workshop grinding wheel
- 9V Battery
- Tongs or pair of pliers
- Heat resistant mat
- Protective wear

Figure 7.1

Safety / Risk assessment

- ★ Wear protective eye wear and clothing and handle the steel wool with tongs /pliers
- ★ Use a sensible size of steel wool and burn it above the heat resistant mat

Here's how

1. Use a lighter and try to light a steel nail. Why doesn't it catch fire?
   Simply because the surface area is too small!

2. Hold the nail to a workshop grinding wheel and see how the friction energy breaks the steel into tiny particles and heats them till they burn white hot. We call the burning particles "sparks" (Figure 7.2). They ignite easily as their surface area/volume ratios are much larger than those of the nail.

$$4Fe + 3O_2 \rightarrow 2Fe_2O_3 + \text{energy}$$

3. Take a wad of fluffed out steel wool with a pair of pliers or metal tongs. Hold it over a heat resistant mat and touch the battery's terminals to the steel wool (Figure 7.3).
   Due to the high resistance in the thin single strands, the wires will heat up. This energy is sufficient to start the oxidation process of the steel wool. The process races along the strands in the wad until all is oxidized. You will be be left with only iron oxide.

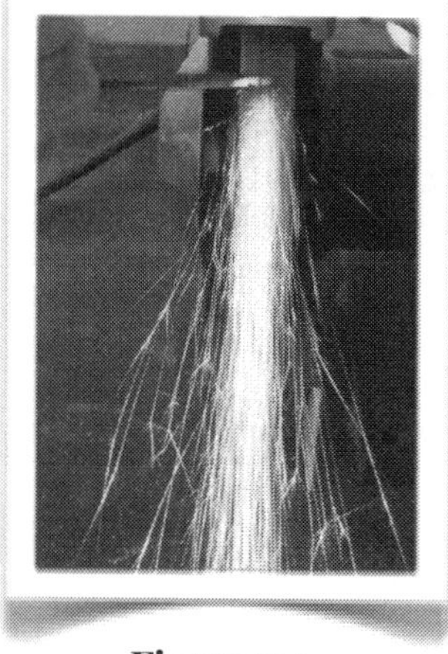

Figure 7.2

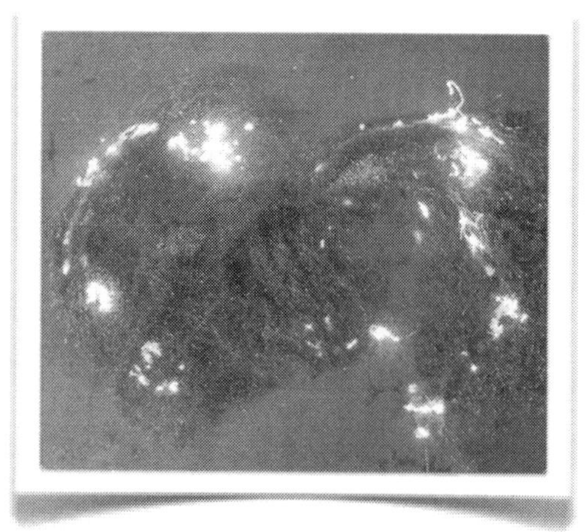

Figure 7.3

4. Fan the combustion by increasing the oxygen 'concentration' by gently blowing on the red steel wool.

Why did this happen?

We have three changed conditions here that greatly increased the rate of combustion:

- An increase in **temperature** brought about by flywheel friction and electrical energy from the battery on a small object. It is much easier to increase the internal energy of smaller objects than that of larger objects with decreased area:volume ratios.
- The large **surface area**. Smaller particles have an increased surface area, which means better exposure to the atmosphere (and oxygen).
- The increased **oxygen supply** rate. By blowing onto the hot steel wool, the rate of oxygen supply is increased. The same experience we have with bush fires being 'fueled' by wind.

Figure 7.4

Another example: When we enjoy the fizzing of a sparkler at a party, we stare at the oxidation of iron and magnesium and are left with a burnt out wire covered in metal oxides (Figure 7.4).

$$Mg + O_2 \rightarrow MgO$$
$$4Fe + 3O_2 \rightarrow 2Fe_2O_3$$

Is this the same as rusting?

Yes and no. Rusting or corrosion is oxidation too. If you leave a steel nail and steel wool outside exposed to the elements for a night or two, you will experience a process of oxidation that occurs over a number of hours. This oxidation only occurs in the presence of water and oxygen. It is an electrochemical process in which different parts of the iron surface act as electrodes in a cell.

$$4Fe + xH_2O + 3O_2 \rightarrow 2Fe_2O_3 \cdot xH_20$$

B. Burning a Metal Pencil Sharpener in Steam [★★★★☆]

The Background

Magnesium

Magnesium is a beautiful silvery metallic element belonging to Group 2 of the periodic table. It is chemically very reactive and bonds easily with oxygen to form a protective magnesium oxide coating. When ignited magnesium burns in oxygen with an intense white flame.

It is used in flares and fireworks and in earlier years photographic flashes were produced by simply blowing a puff of magnesium powder into a flame.

$$2Mg + O_2 \rightarrow 2MgO + energy$$

Magnesium is used extensively as a lightweight alloy usually in conjunction with aluminium in aircraft, bicycles, pencil sharpeners and more.

Many a chemistry student will know that Group 1 alkali metals such as sodium and potassium react violently with cold water. So how does magnesium react with water?

Magnesium powder slowly liberates hydrogen from cold water[1] but burns brilliantly in steam to the oxide and hydrogen.

$$Mg + H_2O \rightarrow MgO + H_2 + energy$$

Burning or molten magnesium metal reacts violently with water. It reduces water to flammable hydrogen gas and forms magnesium hydroxide.

$$Mg + 2H_2O \rightarrow Mg(OH)_2 + H_2 + energy$$

As a result, water cannot be used to extinguish magnesium fires - the hydrogen gas will produce explosive gas mixtures.

Due to the extensive heat, a fourth exothermic reaction shows its face - that between 'inert' nitrogen and magnesium.

$$3Mg + N_2 \rightarrow Mg_3N_2 + energy$$

Furthermore, even carbon dioxide extinguishers should be locked away. A fifth exothermic reaction releases even more amounts of energy.

$$2Mg + CO_2 \rightarrow 2MgO + C + energy$$

It is because of these five exothermic reactions of magnesium with common chemicals (H_2O, O_2, N_2, CO_2) that some of the most catastrophic accidents in motor sports history occurred in conjunction with the use of magnesium as racing car alloy:

- In the 1968 French Grand Prix Jo Schlesser was killed when his newly prototyped Mg-Al alloy Honda racing car left the track and collided with an embankment. The fuel and heat ignited the magnesium alloy

---

[1] This is how magnesium can be identified by simply cleaning it with sandpaper and then placing a drop of water on the surface. Small H_2 bubbles will indicate the presence of Mg.

frame and Schlesser had no chance of survival in the ensuing fire. Fire fighters were ill equipped to handle the magnesium fire.

- Another driver, Pierre Levegh, and 83 spectators were killed at the 1955 Le Mans race when his magnesium bodied car collided and was engulfed in flames. Rescue workers unknowingly used water on the magnesium fire which only intensified the inferno.

Aluminium

Aluminium is a silvery-white metallic element and belongs to Group 13 in the periodic table. Although its common use is in the kitchen for wrapping food, it is surprisingly reactive.

As with magnesium, it is instantly covered by a protective aluminium oxide layer that forms from the oxygen in the air. This layer has excellent resistance to corrosion, is hard and transparent but once removed, exposes aluminium as a very reactive element. Powdered aluminium is used in flash powders and as a rocket fuel.

Aluminium does not react actively with cold or hot water due to the need for huge amounts of activation energy. Amazingly enough, when mixed with magnesium the ensuing magnesium oxidation produces sufficient heat to get the aluminium going too!

Magnesium-Aluminium Alloys

Magnesium in aluminium alloys are used in a range of 0.5 to 98% Mg, the low-magnesium alloys having the best formability and the higher magnesium alloys increased strength. Packaging accounts for 50% of the demand for magnesium/aluminium alloys, the principal use being beverage can manufacturing.

Good quality metal pencil sharpeners are machined from magnesium alloys as fine tolerances are required in sharpeners to hold and shape a pencil point properly. Magnesium is hard enough to be worked with a milling machine at these tolerances, where aluminium is too soft and malleable. Die-cast metal bodies on the other hand cannot be made to this precision[2]. The typical chemical content of a magnesium pencil sharpener by mass is:

Mg 95.5 %; Al 2.8 %; Zn 0.7 %; Mn 0.2 % Ca 0.4%

For us this provides the opportunity to cheaply buy a piece of magnesium and demonstrate magnesium's reactivity in a spectacular fashion.

---

[2] Information supplied by KUM® Company, Germany

Safety / Risk Assessment

This reaction is potentially dangerous but should not, with appropriate precautions taken, be more hazardous than the famous Thermite reaction. It is much easier to put together and quite spectacular. Simply stick to the safety rules:

- ★ Only perform in a fume cupboard or outdoors behind a polycarbonate safety screen
- ★ Use appropriate base heat resistant sheets
- ★ Wear protective clothing, heat-resistant gloves and safety glasses
- ★ Do not touch the beaker or residue with bare hands. Use tongs
- ★ Burning magnesium produces bright light that may cause temporary loss of sight. Instruct students not to look directly at the burning sharpener
- ★ The demonstration produces intense heat and hydrogen gas. A dry-powder fire extinguisher should be readily available at all times. DO NOT use water or a carbon dioxide extinguisher on any magnesium fire
- ★ The near-impossibility of smothering and high temperatures generated make magnesium combustion reactions potentially hazardous. Appropriate precautions must be taken before the sharpener is heated
- ★ Keep flammable material away from the area
- ★ Flasks may be re-used but do not use any cracked glassware
- ★ The audience should be standing well clear of the demonstration

What you will need

- Magnesium alloy pencil sharpener
- Nichrome wire, diameter: 0.30 to 0.45 mm SWG: 31 to 26
- 250 mL Erlenmeyer (conical) glass flask or round bottom boiling flask
- Iron rod (pencil size)
- Retort stand, boss head and clamp
- Metal gauze
- Gas soldering torch / heat gun ➔ ➔ ➔ ➔
- Bunsen burner or gas camping stove
- Small Phillips screwdriver

Figure 7.5

There are three simple ways to determine if a metal pencil sharpener contains **magnesium**:

1. The lower density is an indication. *Geeky!*
2. The reaction of water on its surface when cleaned with sand paper. Tiny hydrogen bubbles form. Check with a magnifying glass. *Cool!*
3. The school boy test. *Wacky!*
   Lick the sharpener lightly with your tongue, touching both the blade and metal body. You should experience a tingling sensation. The taste of a weak electrochemical cell. Here are the cell half reactions:

$$Mg \rightarrow Mg^{2+} + 2e$$
$$2H_2O + 2e \rightarrow H_2 + 2OH^-$$

Here's how

1. Remove the blade of the sharpener with a screwdriver. You may need a small pair of pliers to hold the screwdriver. Cut a 30 cm length of Nichrome wire and fold back to produce a double wire of 15 cm.

2. Insert the wire through the hole in the sharpener. Wind the wire tightly around the body and tie it up tightly (Figure 7.6). This is important as the sharpener has to hang in there as long as possible. Curl the other end of the double wire around the metal rod.

Figure 7.6

3. Add water to the glass flask until it is ¼ full. Position on the gauze above the burner and secure the flask with the retort clamp. Adjust the wire length so the sharpener is suspended about 2 to 3 cm above the water level (Figure 7.7).

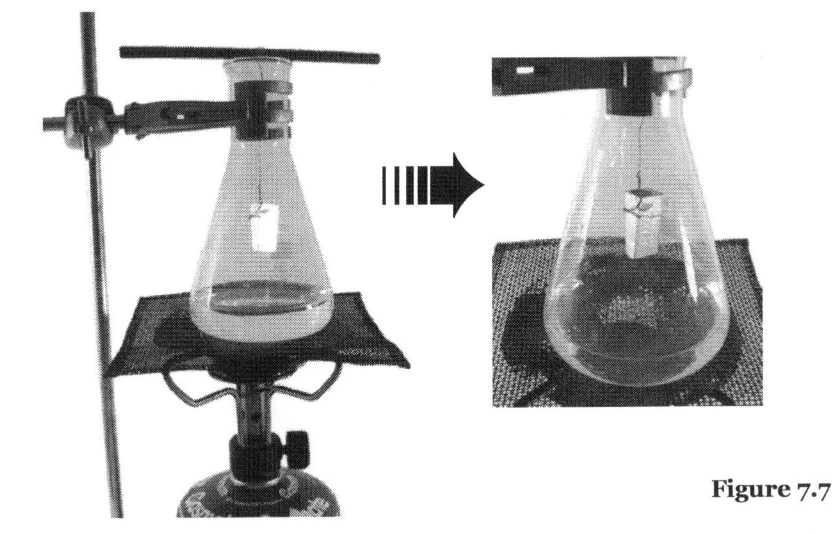

Figure 7.7

4. Heat the water to boiling point and turn down the burner heat slightly once the water starts boiling.

5. Ignite the gas torch, lift the sharpener from the beaker using the rod and strongly heat the sharpener while suspending it above the flask mouth (Figure 7.8). This is the activation energy phase.
   Keep heating, focusing on the bottom part of the sharpener as this is where you want the combustion to start. Heat until it starts shooting a

few **magnesium sparks**. This is the sign to lower the sharpener into the flask, check that the rod is well supported by the flask, **close the gas supply** and move back 3 meters. If after one minute the reaction does not start, approach and re-heat the sharpener.

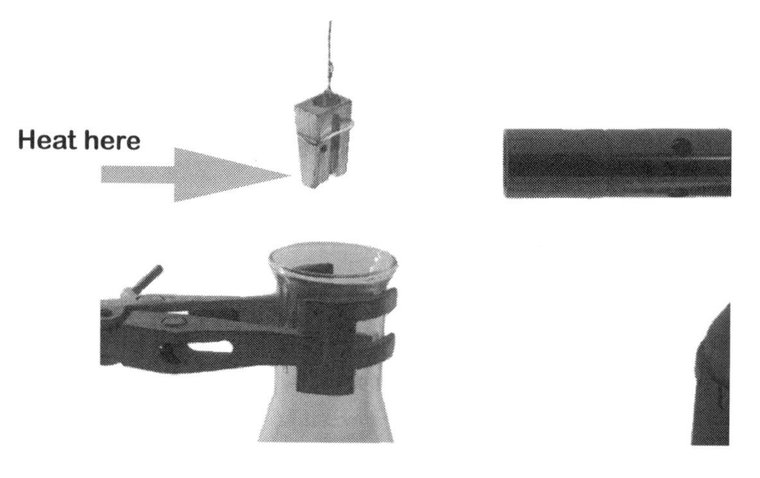

Figure 7.8

6. Once the glowing reaction starts there is **no stopping it**. There will be a quick build-up in intensity as the magnesium and hydrogen burn brightly in the steam above the water. As heat builds up the alloy will melt and drop into the water where the reaction will continue. Do not move closer, the H_2 gas build-up may cause it to explode.

Figure 7.9

7. Only approach the flask once the reaction subsides.
   When all has cooled the residue of oxides and hydroxides can be shown to all. What a remarkable chemical change (Figue 7.10)!

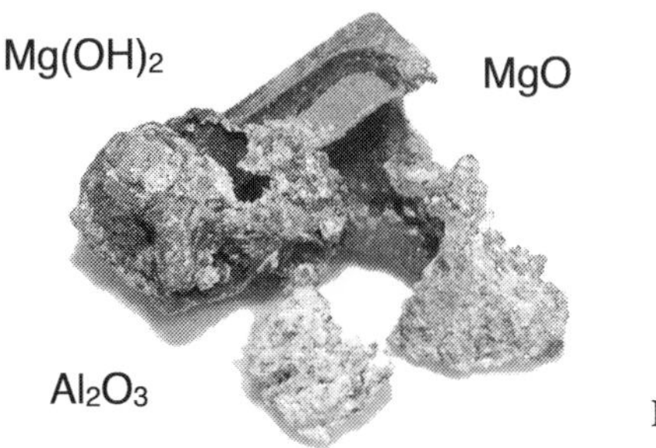

Figure 7.10

Disposal

Allow the solids produced to cool to room temperature. The glass flask sometimes cracks and should then not be re-used. Discard all solids in the waste bin.

C. Drawing electricity from a pencil sharpener

There are two other interesting non-combustion demonstrations that can be done with a magnesium pencil sharpener.

1. Cathodic protection

[☆☆☆☆☆]

When the sharpener comes in contact with an electrolyte (eg. table salt solution), the electrolyte will act as a salt bridge and an electrochemical cell will be established. This electrochemical cell is made up of the sharpener body (Mg) as the anode and the steel (Fe) blade as the cathode:

Anode (oxidation): $Mg \rightarrow Mg^{2+} + 2e^-$

Cathode (reduction): $2H_2O + 2e^- \rightarrow H_2 + 2OH^-$

Bubbles of hydrogen will stream off the steel blade and the salt solution will become more alkaline due to the formation of OH^-.

What you will need

- Magnesium pencil sharpener
- Table salt (NaCl) solution (any strength)
- Small glass beaker
- Universal Indicator

Here's how

1. Make up a table salt solution (any strength) in a small glass beaker. Add two drops of universal indicator - you'll have a neutral green solution.

2. Drop the sharpener in the solution. Hydrogen gas will immediately start bubbling off the steel blade and the metal will turn black. The solution will turn purple → basic.

3. After a few minutes a white gel precipitate ($Mg(OH)_2$, insoluble) will form.

4. Over the next three days the sharpener body will be sacrificed (corroded) and only the blade will be left (Figure 7.11)!

The magnesium sacrifice after 48 hours.

Figure 7.11

The magnesium acts as a **sacrificial anode** and protects the iron blade cathode from corroding. The process is known as **cathodic protection** and is used to protect metal structures such as bridge foundations, pipes, storage tanks, ships, oil well casings and many more.

Teaching Extension Questions

✓ If you started with a sharpener body with mass of 4.0 g (0.17 moles of Mg), calculate the volume of hydrogen to be formed for the complete dissolution of the magnesium.
The answer is 3.8 liters. I have tried this reaction in a soft drink bottle joined to a balloon to collect the H_2 gas but unfortunately the effusion rate through the balloon is faster than the production rate.

✓ Why didn't the steel blade 'rust'?
The steel blade acts as the cathode and is protected from corrosion (oxidation) by the much easier oxidized magnesium - the process of cathodic protection. The blade did not give up any of its atoms as only water was reduced to hydrogen and the hydroxide ion on its surface.

✓ Explain the solution colour change from green to purple.
The OH^- ions formed in the reduction half cell are basic causing the

solution pH to increase with the accompanying universal indicator colour change.

2. A Galvanic Cell

[☆☆☆☆☆]

Can we use the electrical current generated in the electrochemical cell to do useful work?

What you will need

- Magnesium pencil sharpener
- Voltmeter or multimeter
- Digital clock powered by one AA battery
- Crocodile clip wires (2)
- Table salt (NaCl) solution (any strength)
- Cola drink
- Petri dish

Here's how

1. Unscrew the blade from the sharpener body.

2. Connect the crocodile clips to the volt meter probes.

3. Connect the positive wire from the volt meter to the blade and the negative to the magnesium body.

4. Pour the salt solution into a petri dish, dip the electrodes (blade & body) into the solution and record the reading.

5. Now replace the volt meter with a digital clock. Connect the positive wire from the battery compartment to the blade and the negative wire to the sharpener body (Figure 7.12). Does this power the clock?

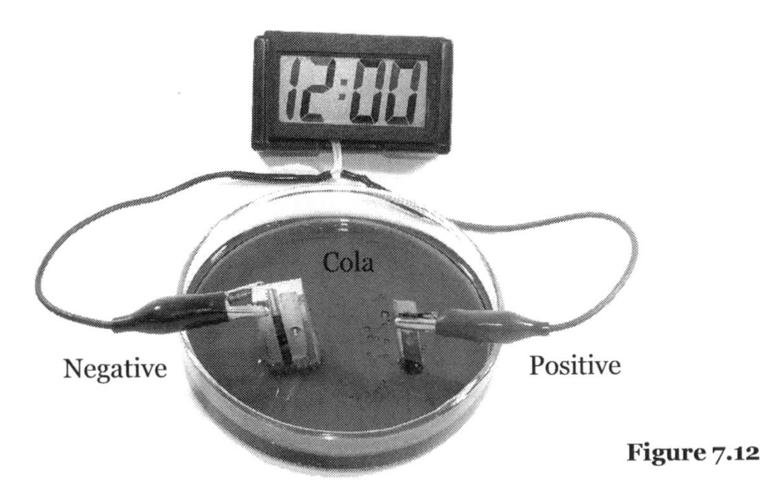

Figure 7.12

6. Repeat, replacing the salt solution with the Cola drink.

How do your results compare with mine?

	Salt solution	Cola drink
Mg/Fe	1.34 V	1.94 V

The magnesium (anode, (-)) and steel blade (cathode, ⊕) acted as the electrodes in a galvanic cell. The Cola performed better as a "salt bridge" as it provided the option of a more spontaneous reduction reaction. Cola contains phosphoric & citric acid, both sources of hydrogen ions (H^+). These ions were reduced to hydrogen gas on the blade's surface:

Anode (oxidation): $Mg \rightarrow Mg^{2+} + 2e$

Cathode (reduction): $2H^+ + 2e \rightarrow H_2$

The galvanic cell produces a minute current but two of these cells in series produce sufficient voltage and current to light a LED.

If MacGyver were still around he would be able to escape from jail using a burning magnesium sharpener in steam as well as powering his mobile phone from six sharpeners in Cola!

Key terms

Magnesium, alloys, magnesium reactivity, exothermic reactions, combustion, oxidation, redox reactions, chemical reaction rates, electrochemistry, cathodic protection, electrochemical cells, galvanic cells

References

1. Brink, J. A., Missiles & Pencil sharpeners. IMSTUS News, University of Stellenbosch, 1985, p 11 - 12
2. Isaacs, A, Dictionary of Science, Grange Books, London, 2005
3. Shakhashiri, B. Z., Chemical Demonstrations: A handbook for teachers of chemistry, Vol. 1, The University of Wisconsin Press: Wisconsin, 1983; p 96 - 98
4. Karukstis, K.K., Van Hecke, G.R., Chemical Connections The chemical basis of everyday phenomena, Harcourt Academic Press, 2000
5. Selinger, B., Chemistry in the Marketplace, Harcourt Brace Jovanovich, Sydney, 1998
6. Gillespie, R. J., Humphreys, D. A., Baird, N. C., Robinson, E. A., Chemistry, Allyn and Bacon, Massachusetts, 1986

7 Sugar can go Bang!

The Dust Explosion Pentagon

The solid with the lowest bulk density on Earth relies fully on its surface area for its extraordinary insulation & structural capabilities. Aerogel (fig. 8.1), also known as "frozen smoke" is 99.8% air and boasts a huge specific surface area of 900 m^2/g.

Surfaces of substances determine their chemical activity and ability as this is where chemical interactions happen. Studies of catalysts, absorbents, surfactants and corrosion substrates all focus entirely on the surfaces of the substances.

In this chapter we will demonstrate the relationship between surface area and reaction rate, but first we will look at the reason why 'smaller' in chemistry means 'larger'.

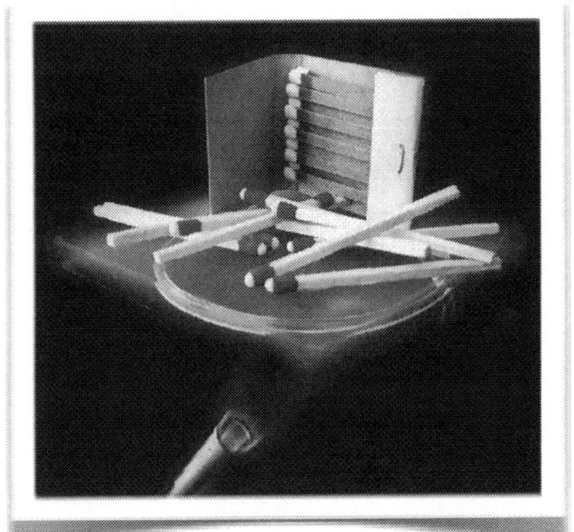

Figure 8.1
Aerogel's remarkable insulation capabilities
Photo: NASA

The Background
Smaller means larger

In chapter 6 we demonstrated the remarkable effect particle size has on the ignition and oxidation of steel. Rocket scientists will also attest to the importance of getting the fuel introduced in the combustion process in the smallest possible particle size.

Even around the house there are many reactions in which a smaller particle size can bring substantial change in the reaction rate. Let's have a look at a few we encounter everyday:

- We get up in the morning and pop an effervescent tablet into a glass of water. By breaking the tablet into smaller pieces, we can accelerate the dissolution process
- We chew our food so it digests quicker
- We start a fire in the fireplace and choose the finer sticks to get the fire started. Later we add the larger logs
- We bake a cake and put the flour through a sieve
- When having a cup of tea we choose granular sugar over sugar cubes as it dissolves quicker
- In the stove's fume extractor we have finely divided activated charcoal with a surface area of 800 m^2 per gram to absorb as many molecules as possible

So smaller seems to always produce a faster result or increased reaction tempo. Why? Let's do a simple calculation:

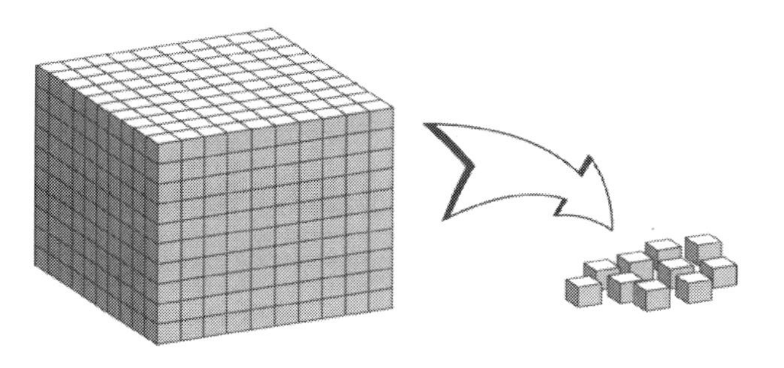

Figure 8.2

In order to start a fire, we use a wooden cube with 10 cm sides (Fig. 8.2). The area that is exposed to air is: 10 x 10 x 6 = 600 cm^2
If we chop it up into cubes with 1 cm sides (figure), the total exposed surface area becomes: 1 x 1 x 6 x 1000 = 6,000 cm^2
Further subdivision of the 1 cm cubes into 1 mm cubes provides a total surface area of 6,000,000 cm^2
So we get a **10,000 times increase in surface area** by simply dividing the block into smaller parts. This brings many more wood (cellulose) molecules in contact with oxygen molecules and explains why we usually start fires with twigs and then add the logs.

Still, many students find it hard to grasp the fact that tiny dust particles produce in total a much larger surface area than a solid block of the same compound. For them, I have a second method that would hopefully convince the doubtful.

Click four Lego® blocks together (Figure 8.3). Have students measure and calculate the surface area. Then

Figure 8.3

ask them to break the four blocks apart and calculate the surface area of all blocks combined. That should convince them.

Smaller in size means larger in surface area and hence, larger in reaction rate.

In the following demonstration, fine **sugar dust** will be dispersed in **air** and ignited by a flame. For such a dust explosion to occur, the presence of five conditions are essential - also known as the **Dust Explosion Pentagon**. They are:

- Combustible dust
- Dispersion of the dust into a cloud (in sufficient concentration)
- Confinement
- Oxidant (usually air)
- Ignition source

If any one factor is missing then there can be no explosion. Note that the well-known Fire Triangle incorporates only three of these conditions.

Because of the hugely increased surface areas, the **ignition temperatures** for dust explosions are relatively low and this adds to the adverse dangers these reactions hold. To ignite sugar dust in a 1 m^3 vessel, a temperature of only 370°C (698°F) is required at a dust concentration as low as 60 g/m^3. And the resultant explosion pressure can produce pressures of up to 9.5 bars (137 psi).

Can we pull this off in a coffee / Milo® can?

Sugar Dust Explosion

[★★★☆☆]

Safety / Risk assessment

- Do not use open flames in the proximity of flammable gases
- Have a fire extinguisher readily available
- Check that you are a safe distance from flammable material such as paper, curtains, etc.
- Check that you have sufficient head space as the lid will be propelled up
- Wear safety glasses and a lab coat
- Spectators should stand at least 4 m off and warn them about the lid that can be deflected from the ceiling
- There is no need for ear protection or safety glasses for the audience as no shrapnel flies out

Figure 8.4
Tubing, tea-light & dusting sugar

What you will need

- Large coffee / Milo® can
- Plastic tubing, ~ ID 6 mm, ~ OD 8 mm, length 1.5 meter
- Tea-light candle (candle with a wide base in metal container)
- Toy balloon pump to fit the tubing
- Plastic egg cup *or* disposable wine glass *or* small plastic cup
- Epoxy putty
- Fire lighter with long stem
- Drill and drill bit, diameter of tube
- Gas torch
- Metal file
- Fine **organic powder** such as lycopodium or icing sugar: Lycopodium powder (aka Dragon's Breath at magic shops) is our first choice but has become very expensive and can be hard to find. I have tested many maize based products in my search for a supermarket substitute and have found **Dusting Sugar** a great alternative for lycopodium. In Australia it is branded as "Vanilla Bean Dusting Sugar" and available in most supermarkets. It contains icing sugar, cornstarch and vanilla bean powder. This mixture produces great results - much better than pure cornstarch or icing sugar (confectioners sugar). But in the end it is the particle size that matters.

Here's how

1. Clean out the metal can and drill a hole about 1 cm (0.4") from the base in the side wall. The diameter should be that of the plastic tube. Clean all burrs with a metal file. (You may prefer to use long nose pliers to widen the hole).

2. Wear heat protective gloves and heat the back end of the drill bit or a metal rod in a bunsen flame. (I use the drill bit as its diameter is the same as that of the tube). Melt a hole in the bottom of the small plastic cup (Figure 8.5).

3. Push the tube through the metal can wall and through the hole in the plastic cup. It can protrude a little bit into the cup. *Leave enough tubing inside the can so you can easily lift the cup and tubing out to inspect and clean (Figure 8.7).*

Figure 8.5

4. The next step is optional: Mix some epoxy putty components well and seal the tube hole on both sides of the can wall. Leave to cure.

5. The balloon pump fits onto the other end of the tube (Figure 8.6). Extend its handle so it is filled with air.

Figure 8.6
Pump, tube
& can set-up

6. Place the candle alongside the cup inside the can and scoop **two teaspoons** of the sugar into the cup. Now we are ready for action.

7. Light the candle with a long stemmed gas lighter.

8. Check that all spectators are standing at least 4 m (13') off.

9. Push the lid **tightly** onto the can.

10. Do the countdown and pump the air in one brisk thrust down the tube (Figure 8.7).
    Expect a BIG ball of fire as the lid flies to the ceiling.

Teaching Extension

Place a teaspoon of the powder on a heat resistant surface and try to set it alight with a lighter. No success!

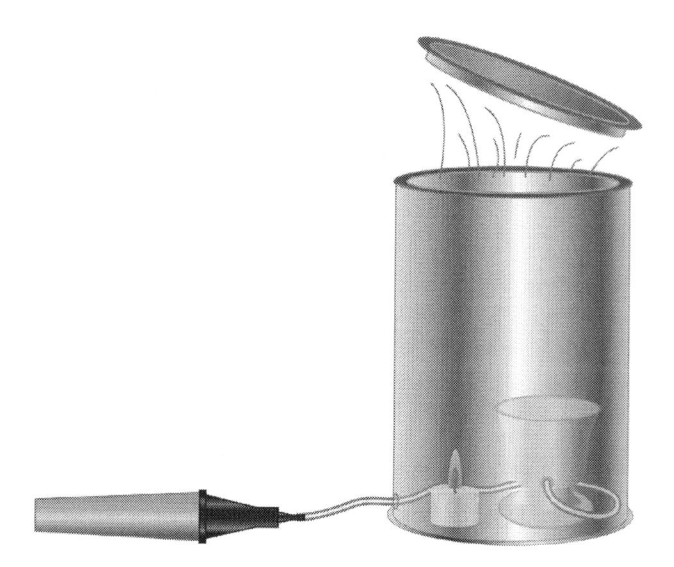

Figure 8.7

Discussion

In this demonstration you have fulfilled the requirements for the Dust Explosion Pentagon: You have **dispersed** fine particles (large surface area) of a **combustible substance** in **oxygen** (air). The candle flame provided **activation energy** to get the ball rolling and all was neatly **contained** inside the tin. You have increased the concentration (amount) of reagents that reacted by exposing larger surface areas in blowing the sugar particles into the air. This is the combustion reaction:

$$\underset{\text{sucrose}}{C_{12}H_{22}O_{11}} + 12O_2 \rightarrow 12CO_2\,(g) + 11H_2O\,(g) + \text{energy}$$

Figure 8.8
Rod Kirkwood of Horsham College puts lycopodium powder to the test. Here we used a large milk can with a cork lid. It is safe to blow down the tube as no flash-back is present.
Photo: Michael Rudolph, The Wimmera Mail-Times

More reactants reacted within a shorter time and a huge gas increase (11 moles per mole of sucrose) of combustion products followed. The lid just had to pop!

Real life hazards with dust explosions

For many a worker the concept of dust explosions is a real threat and form part of their daily risk assessment. The occurrence of dust explosions are hard to avoid in many processes where combustible powders are handled. Coal dust explosions are well-known but it is the lesser known more mundane materials such as grain, flour, sugar, starch, milk powder, rubber and sawdust that ever so often prove that Chemistry has the last say.

In coal mines, a methane explosion can initiate a coal dust explosion which can engulf the entire mine in seconds by setting up secondary dust explosions. The initial primary explosion causes a pressure wave that dislodges settled dust that may produce a much larger secondary dust explosion. A very real example is the tragedy in which 29 miners died in the November 2010 Pike River Coal Mine in New Zealand.

But if working in a manufacturing plant sounds safer, consider this:

- In January 2003 polyethylene dust ignited in the ceiling space of a pharmaceutical company in North Carolina, USA, killing six workers and injuring thirty eight. The explosion was so intense it could be heard forty kilometres away
- In February 2003 a series of secondary dust explosions ripped through a Kentucky acoustics insulation plant fatally injuring seven workers
- On February 7, 2008 a huge explosion and fire occurred at the Imperial Sugar Refinery in Georgia, USA causing fourteen deaths and seriously injuring thirty eight others. The explosion was caused by accumulated sugar dust in the packaging facility
- The US Chemical Safety Board (CSB) reports that between 1980 and 2005, 281 industrial combustible dust explosions occurred in the USA.

Key Terms

Reaction surface area, reaction rate, activation energy, exothermic reactions, dust explosions, deflagration

References

1. Selinger, B., Chemistry in the Marketplace, Harcourt Brace Jovanovich, Sydney, 1998
2. Barton, J., Dust explosion prevention and protection: A practical guide, Institute of Chemical Engineers, 2001
3. See USA dust explosion reports and educational videos on the CSB website: www.csb.gov/investigations.aspx

8 Sugar can go Whoosh!

Sugar-Based Rocket Fuel

There is hardly an amateur science activity that gives as much satisfaction as the 'whoosh' launch of a home-made rocket piercing the sky!

In the previous chapter, we utilized sugar's surface area and now we will blend it with an oxidizer to speed up its combustion rate. We are out to prove that the chemical potential energy contained in the molecules of 3.5 g sugar can drive a tiny rocket to heights of almost 80 metres.

The Background

Rocket propellants develop a thrust by producing copious amounts of hot gases that are propelled away from the rocket body. This action alone does not propel the rocket, but the reactive force described by the Third Law of Newton, that acts as an equal and opposite thrust on the rocket body.

The amount of thrust developed by a rocket engine depends, amongst other factors, mainly on

- the **velocity** with which the burning gases leave the chamber and
- the **mass** of the burnt expelled gases (F = ma). *The Space Shuttle's two solid rocket boosters weigh more than half a million kilograms!*

In order to produce hot expanding gases, propellants need to be exothermic (heat generating), act independently of the surroundings and produce the gas at a constant high rate.

Most rocket fuels are made up of at least two components that act in a redox reaction: The oxidizer and the fuel. The role of the oxidizer is that of supplying oxygen to the fuel to make the combustion self-contained, self-sustained and non-reliant on surrounding gases. A great rocket mix should be able to burn quickly in the complete absence of air oxygen.

There are hundreds of different fuel / oxidizer mixing options. Here are five that have proven to be very popular in model rocketry:

Fuel	Oxidizer
Sulphur	Potassium chlorate
Magnesium	Potassium chlorate
Powdered sugar	Potassium permanganate
Charcoal & Sulphur	Potassium nitrate
Aluminium dust	Ammonium perchlorate

Table 9.1

The Estes™ type of model rocket engines use black powder as a propellant. Their composition comprise of 71.8% Potassium nitrate, 13.5% Sulfur, 13.8% Charcoal and 0.95% Dextrin.

Preparing any of these mixes will require previous pyrotechnic mixing experience and special safety equipment. Furthermore, the chemical instability of some of these compounds are to be taken very seriously! In this book we are focusing on safe, spectacular activities, so we will not engage in mixing any of the more risky explosive-prone compounds. We love our fingers and are after simple, easy methods.

This is our challenge:
Prepare a simple, safe rocket that is easily reproducible.

In this project we will utilize a **sugar** as fuel and **potassium nitrate** as oxidizer.

Sucrose, $C_{12}H_{22}O_{11}$
Sucrose is comprised of one molecule of glucose linked to a fructose molecule. It occurs widely in plants and is particular abundant in sugar cane from which it is extracted and refined for table sugar. At 200°C (392°F) it starts decomposing (caramelizing).

Potassium nitrate
Potassium nitrate (KNO_3) also known as saltpetre, is a well-known oxidizing agent and considered very stable and safe to work with. It is the

oxidizer in gunpowder but also found in salami and meats as a preservative as well as in de-sensitizing toothpaste!

My first knowledge of potassium nitrate came about when using it as a food preserving agent. In South Africa, my country of birth, it is the chemical of choice for preserving air-dried meat known as 'biltong' in a process very similar to that of preparing traditional 'jerky'. I am including a photo at the end of this chapter to whet your appetite (Figure 9.15).

Back to lab chemistry. Potassium nitrate decomposes above 400°C (752 °F) to produce oxygen and potassium nitrite. This is very handy as it acts as a source of internal oxygen.

$$2KNO_3 \rightarrow 2KNO_2 + O_2$$

In contrast with powerful oxidizers such as chlorates that give up all their oxygen when decomposing,

$$2KClO_3 \rightarrow 2KCl + 3O_2$$

potassium nitrate only hands over one third its oxygen. It is therefore a slower but safer oxidant to use. It also explains why our propellant will be classified as a "low impulse, slow burning" propellant. Serious rocketeers classify this as a beginner's composition and have likely named it "candy propellant".

Determining the actual **combustion equation** is probably the most complex step in rocket engine analysis. Increased temperature and pressure amongst other factors play a major role in determining what the real end products and their relative ratios will be. Richard Nakka (Ref. 1) and other developers have made huge contributions over the years in developing software that does the more complex iterative calculations required. Using their calculations, the following equation is derived as the best representation of what the true combustion ratios are at a pressure of 68 atmospheres - the pressure attained inside a medium sized model rocket:

$$C_{12}H_{22}O_{11} + 6.3KNO_3 \rightarrow 3.8CO_2 + 5.2\,CO + 7.8H_2O + 3.1H_2 + 3.1N_2 + 3.0K_2CO_3 + 0.3KOH$$

When calculated as relative masses the reagent ratio becomes: 65 : 35 in $KNO_3 : C_{12}H_{22}O_{11}$ The combustion temperature is somewhere between 1,150°C and 1,450°C (~ 2,370°F).

A. Build & Launch a Sugar-Based Rocket

[★★★☆☆]

Safety / Risk Assessment

- ★ Many States have strict **regulations** on the preparation and use of pyrotechnics, so please check with the local authorities before attempting this project
- ★ You may be living in an area that is prone to **bush fires**. Please be considerate and comply fully to all fire protection regulations. Do not fly rockets if any fire hazards exist
- ★ Although this propellant is very benign when compared to other rocket propellants, it is a **rocket propellant** and as such flammable. Once blended, keep it away from all possible ignition sources. Its auto-ignition temperature is estimated at >300°C (572°F)
- ★ Never **grind** the oxidizing and reducing agent (fuel) together!
- ★ Only prepare **small batches** of the propellant at a time
- ★ Always wear at least **safety glasses** (full face cover is recommended in case of a flare-up), a lab coat and heat protective gloves when working with the propellant
- ★ Have a fire extinguisher readily available as well as a large container filled with **cold water**
- ★ Do the drilling through the nozzle epoxy and propellant with a slow speed **bench drill press**

What you will need

- **Potassium nitrate** (KNO_3) or **saltpetre**. *It is used for pickling meats, in hydroponics as fertilizer and in toothpaste for 'sensitive teeth'. This can be bought as a powdered solid off eBay in most countries or from a hydroponics store. In North America, try Lowe's or The Home Depot where it is sold as 'stump remover'. Check the fine print on the container*
- Fine powdered **icing sugar** (also known as confectioners sugar). This is table sugar (sucrose) in a very fine form - large surface area! It is available from all supermarkets
- Round wooden dowel stick, diameter. 9 to 10 mm (0.4"), 12 cm (5")long ("The dowel")
- Small plastic tub with lid for mixing dry propellant
- Balance / electronic scale to 0.1 g
- Copy paper, A4 or US letter (8½" x 11")
- White paper glue
- Epoxy putty (available as a 2 component putty that needs mixing) You may substitute this with *bentonite clay.* This can be found at ceramics (pottery) suppliers but it is also sold as clumping cat litter. This will need some grinding into fine powder. Check that it says 'bentonite' or 'clumping litter'
- Bench drill press and drill bit, 3.5 mm (0.14") diameter

- Glue gun or masking tape
- Wooden skewer sticks
- Thin metal tube used as launching guide
- Sparkler (used as fuse) & lighter **or**
  nichrome resistance wire, speaker wire and 9V battery **or**
  commercial hobby model rocket igniters with battery

Preparing the dowel stick:
Sand the ends of the dowel stick with sand paper and smooth the sharp edges. (Figure 9.1). Mark one end as "hammer". This is the end to which you will apply the hammer.

Smooth

Figure 9.1

1. How to Make the Rocket Body

The rocket design

As mentioned before, our propellant is a low-impulse type and in order to create enough thrust for a lift-off, we need to ignite **as much of the propellant as possible, as quick as possible**. This can only be facilitated by introducing a **hollow core** through 80% of the propellant charge (Figure 9.2).

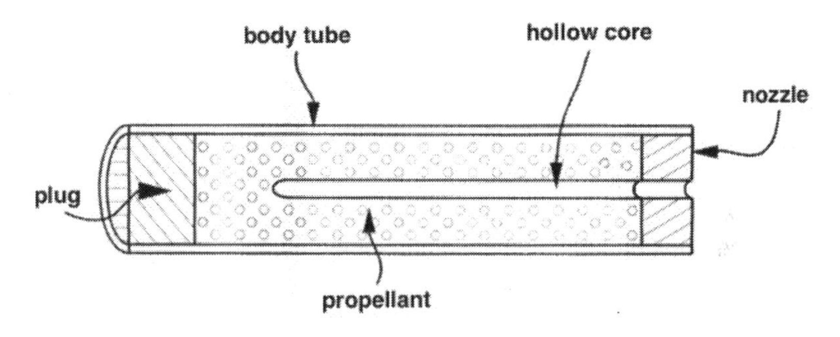

Figure 9.2: Cross-section through rocket body

We could cast or compact the propellant with a hollow core by inserting a coring tool into the body casing, but that asks for more sophisticated tooling and preparations. There is a much simpler way . . .

If you have cardboard tubes available with an inner diameter of approximately 9 to 10 mm (0.4") and the dowel fitting inside them, then you can cut them to size (≈ 70 mm long (2.8")) and follow from point (e):

a. The rocket body tube is prepared from ordinary white paper and bonded with white paper glue.

b. Cut one sheet into three strips of paper, 70 mm wide and 297 mm in length - the length of the A4 paper (Figure 9.3). (If you are using US Letter size, make the strips 2.8" wide and 11" long). Draw a line 50 mm (2") from the one end. Each sheet thus produces three body tubes.

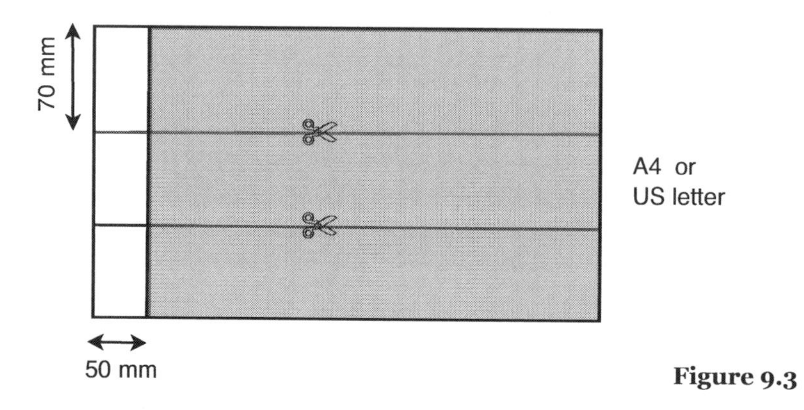

Figure 9.3

c. Cover the grey area on each strip with a thin layer of white paper glue. An icy-pole / ice-cream stick is handy here.

d. Place the 10 mm dowel on the glue-less end and roll the paper tightly around the dowel to form a paper tube (Figure 9.4). Finish off the outside of the tube with a layer of white glue. Put aside for 2 hours to cure.

Figure 9.4

e. When the body tubes are dry, insert a plug into each tube:
   - Cut a small piece or approximately 2 g from the epoxy putty and mix well. Insert the mix into one end of the body tube and place this end down on a piece of waxed baking paper or smooth hard surface;
   - Wet the hammer-end of the dowel stick and insert it down into the body tube. Compact the putty into position by pressing down on the dowel stick (Figure 9.5).

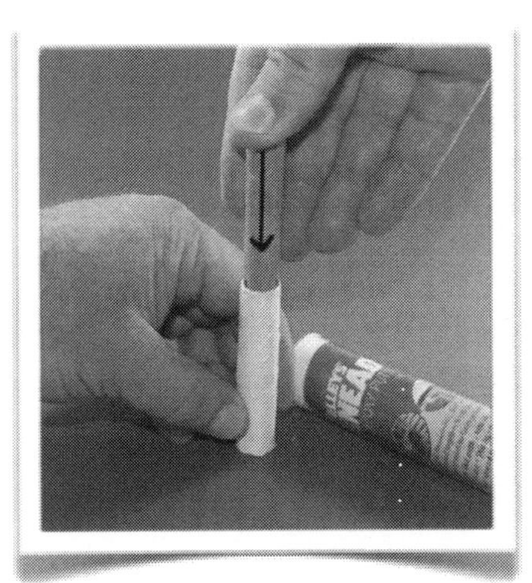

Figure 9.5

- Remove the dowel slowly from the tube so the putty doesn't get pulled along. Finish off the outside putty with a wet finger or dowel. Leave to cure for 30 minutes.
  Alternatively you can simply ram dry *bentonite clay* or two sheets of *toilet paper* (yes, it works!) down the tube with the dowel and hammer. Do this on a solid concrete floor. Use the same procedure as we will use to load the propellant - see next pages. The plug should be ~ 5 mm (0.2") thick.
- Mark this plug with a "T" (for "Top") on the body tube.

2. How to Prepare and Compact the Propellant

We will prepare a rocket using "classic" sugar-based rocket propellant comprising of a mixture of **potassium nitrate** (KNO_3, saltpetre) serving as the oxidizer, and **sucrose** ($C_{12}H_{22}O_{11}$, icing/confectioners sugar) serving as the fuel and binder. Although not a high-performance propellant, its main advantage is the relative ease and safety of preparation and common availability of the ingredients.

Icing sugar contains a low percentage of corn starch $(C_6H_{10}O_5)_n$ but this is acceptable as it is chemically very similar to sucrose. Granular or table sugar may be used also but asks for some rigorous grinding.

The particle sizes of both chemicals are very important as these determine the impulse and burning rate and thus the success of your project. I have found invariably that icing sugar particles are fine enough but the saltpetre usually requires some grinding.

a. **Preparing the chemicals:**
   Grind the saltpetre (about 25 g) in two batches in a **mortar & pestle**. Another way of doing this is to use an empty **wine bottle** and grind a small quantity at a time with a rolling action on a clean, solid surface

for two minutes. The finer you can get this, the better the burning rate will be and the higher the impulse. An electric coffee or herb **grinder** will do the job too in less time with less human effort! Do not use your kitchen grinder - for a few dollars one of these makes a great investment for the lab (Figure 9.6).

Figure 9.6

b. Do the same with the sugar (15 g) if it is not in a very fine form.
Safety: Do not use the same grinder as used for the saltpetre. In pyrotechnics the rule of survival is: Never grind oxidant and fuel together.

c. We will use a ratio of **65% KNO_3** and **35% Sucrose**, a mix ratio favoured by many beginner rocketeers. Each rocket will require around 10 g propellant. So I would suggest a mix of **30 grams** to start with for making 2 rockets. Thus, at the above ratio you will require:
KNO3 19.5 g
Sucrose 10.5 g
Weigh these amounts on an accurate scale into two separate plastic cups.

d. **Blending the two components:**
Transfer the weighed components to a plastic container, seal and shake for **two minutes**. Complete mixing of the components is necessary for optimum and consistent performance. Adding two glass marbles will enhance the mixing and perform further pulverising. Always keep the lid of the container sealed as the propellant is slightly *hygroscopic* and will absorb air moisture.

Safety: Once combined, the two components form a flame- and spark sensitive propellant but is not impact sensitive. So shaking it holds no danger, but be sensible and keep it away from any possible ignition source. (Try this: Wear safety equipment and place a small quantity of the blended propellant on a solid surface and hit it with a heavy hammer - no reaction. Now strike a match and ignite the mix).

e. **Loading the propellant:**
Use another sheet of paper to prepare a small powder scoop as in Figure 9.7:

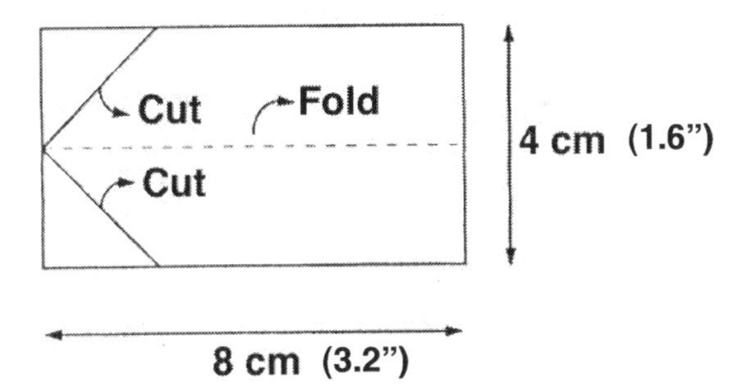

Figure 9.7

f. Use the prepared body tube with hardened plug. Stand the body tube on its top end on a solid base, such as a wooden block on a concrete floor. Or work directly on the floor. With the powder scoop take about 2 mL of the propellant and load this down the tube.

g. Insert the dowel stick and hold the rocket body firmly in one hand perpendicular to the floor. Compact the propellant with three light taps to the stick with the hammer - do not hit too hard as you could damage the tube (Figure 9.8). Keep on repeating this process with small 2 ml powder increments and fill the tube up to **10 mm (0.4") below the rim** of the tube. This space is required for the nozzle plug (Figure 9.2). You may have to steady the dowel with your hand as the tube fills up. Remove any excess propellant from the top inside walls with a paper towel.

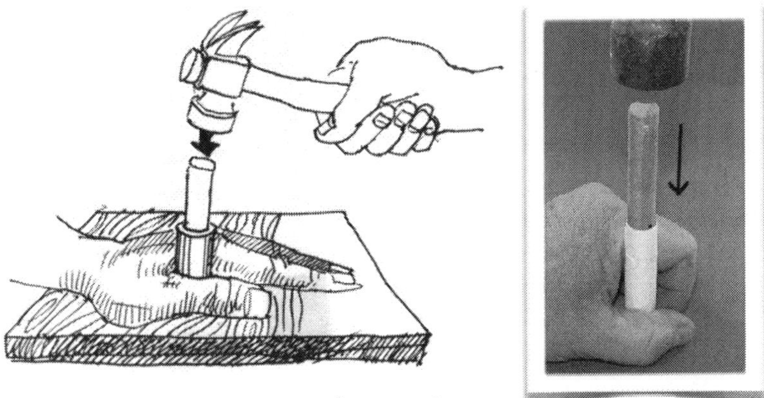

Figure 9.8

h. As before with the end plug, cut a small piece of epoxy putty and mix well with your fingers. Insert this into the open end of the rocket to form a nozzle plug and finish it off with a wetted dowel. Check that the putty adheres well to the inner tube walls. Put the rocket aside for 30 minutes to cure.

i. **Drill the nozzle hole:**
When cured, use a **3.5 mm** (0.14") drill bit and carefully drill through the **nozzle plug** into the propellant to form a central **hollow core**. You will have to drill down to 80% of the propellant depth (Figure 9.2). That is about 45 mm (1.8") for our rocket. Set the drill level so you do not accidentally drill through the top plug. To assist with drilling accurately and consistently into the propellant you'll need a low speed **floor** or **bench drill press** (Figure 9.9). It is indispensable here. Do the drilling stepwise and remove the drilled residue as you go.

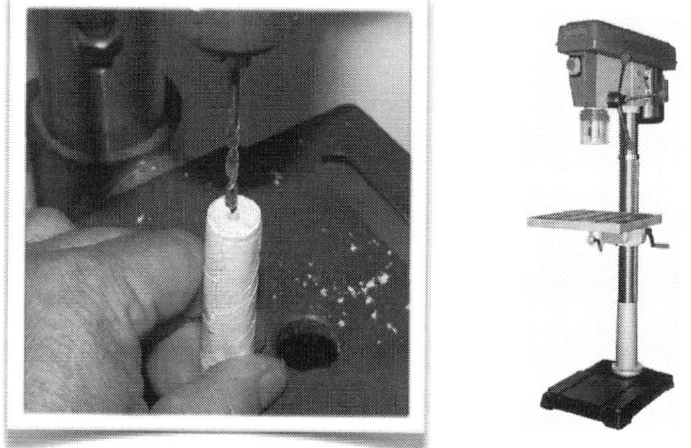

Figure 9.9

3. How to Stabilize & Launch the Rocket

A. Stabilizing the rocket

A rocket body (engine) will never fly on its own as it is aerodynamically unstable. This is true for an inflated balloon too - it has an irregular flight path. Its flight stability will increase only when the "rocket's" centre of gravity (CG) is positioned ahead of the centre of pressure (CP). As they say in model rocketry: "G before P"

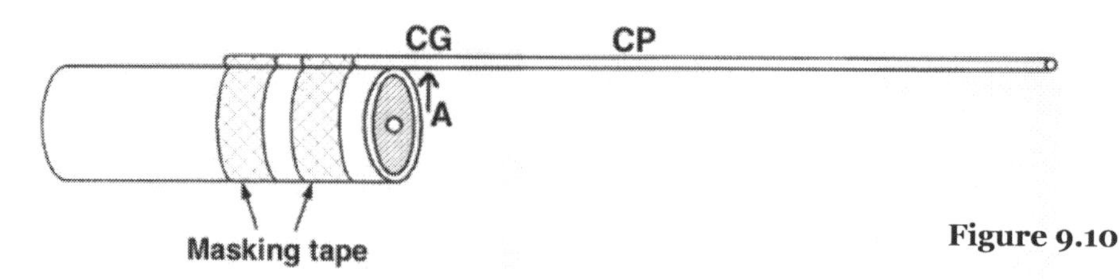

Figure 9.10

Centre of gravity: The imaginary point at which a body's mass seems to be concentrated, around which a free rotating body will spin in space. Also known as the balance point.
Centre of pressure: The imaginary point in a flying body on which all air pressure forces are acting.

So how do we go about stabilizing the 'rocket'?

We simply need to glue or tape a stick to it. This will push its CP **downwards** away from the CG.
Tape or glue a thin dowel stick or skewer stick to the body using masking tape or a glue gun. Balance the rocket at **point A** (Figure 9.10) with a pen held horizontally. Correct the balance by glueing another piece to the skewer stick or cutting a piece from it.

B. Igniting the rocket - finding a fuse

In many states in Australia and the USA, the making and possession of igniter fuse by unlicensed citizens is illegal. So once again we have to find a creative alternative:

- You can make a simple electric igniter using nichrome wire, copper and flex wire and a 9V battery (Figure 9.12), or
- buy commercial model rocketry igniters from a hobby shop, or
- if you have pyrotechnic fuse - by all means use it. Fold one end of the fuse back before inserting it so it forms a hook to prevent it from sliding out. Don't skimp on the length of the fuse.
- But there is an easier way. Off to the party shop this time . . . and buy a few packs of sparklers. (They're not banned!) Cut the bare metal wire (handle) from the sparkler with a pair of pliers. You will be left with the grey pyro part which is tapered on one end. This thinner section can easily be inserted into the rocket nozzle and will "auto-grip" in a 3.5 mm hole (Figure 11). On ignition, the gases will simply expel the sparkler from the nozzle hole.

Figure 9.11
Sparkler as fuse

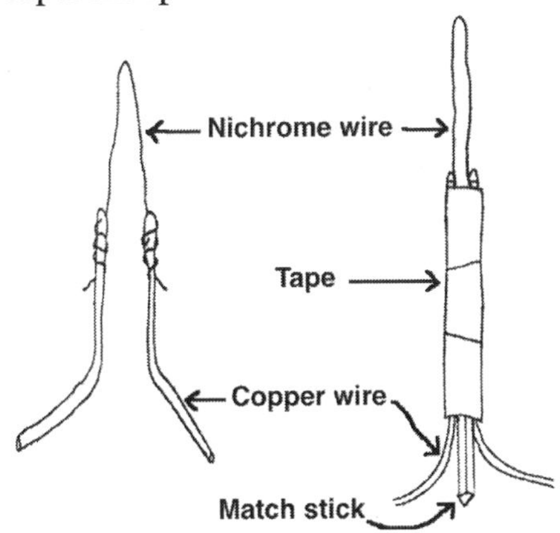

Figure 9.12

Safety / Risk Assessment

- *Never launch a rocket closer than 500m from power lines, hay stacks or a thatched roof house*
- *Do not launch in a built-up area - the rocket has no parachute*
- *If you are close to an airport then you may have to get permission to launch. Check your local council regulations*
- *Rockets should only be launched vertically*
- *Do not launch even if a mild risk of bushfire exists or if the wind is blowing*
- *Spectators should be at least 10 m off and you have to step back 5 m after igniting the fuse*
- *When you have a failed launch, wait at least 1 minute before approaching the rocket*
- *Always wear eye protection*

C. The launch

A thin metal tube is planted upright at the launching site. The rocket's thin dowel stick should slide loosely into this tube. Insert the sparkler into the rocket engine hollow core as far as it will go. It should stay into position itself.
Check that all is in sync with the safety regulations, light the sparkler and step back 5 metres

5 ... 4 ... 3 ... 2 ... 1 ... launch!

The rocket produces copious amounts of smoke leaving a visible trail (Figure 9.13).

Figure 9.13: A static rocket motor test

Disposal

Retrieve the burnt-out rocket body and discard it in a bin when cooled down. The propellant mix can be safely stored for months as both chemicals are very stable. But seal the container well to keep the air moisture out. If you want to, dissolve the unused propellant in water and wash it down the drain.

Teaching Extension

The Reaction Products

Here are some interesting questions related to the rocket's combustion reaction. Have students study the reaction

$$C_{12}H_{22}O_{11} + 6.3KNO_3 \rightarrow 3.8CO_2 + 5.2\,CO + 7.8H_2O + 3.1H_2 + 3.1N_2 + 3.0K_2CO_3 + 0.3KOH$$

and then ask . . .

- Why does the rocket produce a purple (lilac) coloured flame? Burn a small quantity of the propellant in a fume cupboard to see the colour.
  (Potassium flame)

- What are the names of the compounds that showed up as smoke?
  (K_2CO_3 & KOH)

- Are the smoke compound(s) gases or solids?
  (Solids)

- What happened to the CO_2 ; CO ; H_2O ; H_2 ; N_2 ? Are they visible?
  (They are gases and are invisible just like the surrounding air)

- Is there an increase in entropy (ΔS)?
  (Huge increase as 23 moles of gas are formed from 1 mole sucrose)

- Is this an endothermic or exothermic reaction?
  (Exothermic)

- Which of the reaction products would cause pollution concerns if this propellant were to be used to propel commercial rockets?
  (Green house gas CO_2 & caustic KOH)

- What was the *real origin* of the energy released in the propulsion reaction?
  (The sun. The surplus exothermic energy produced in this reaction originated from the rearrangement of the bonds in the sugar and nitrate molecules. The sugar chemical potential bond energies were established during the process of photosynthesis in the sugar cane, powered by solar energy. This is true for all bio- and dino-fuels)

B. Prepare Fire Paper

[★☆☆☆☆]

If sucrose can act as a fuel, how about a similar organic product such as cellulose?
To answer this, we can try an old chemistry trick that links up nicely with saltpetre and its use as an oxidant. You may have seen it before - a blank piece of paper burns in a predetermined pattern along an invisible line.

Prepare the paper

1. Dissolve 2 teaspoons of potassium nitrate in 50 mL (1.7 oz) water. If all solids do not dissolve - no problem.
2. Take a sheet of paper or news paper (preferable) and write your name on it with a glass rod or a cotton ear bud dipped in the saturated solution. Use continuous (cursive) writing and link the name and surname with a line. Use the liquid liberally and mark a spot where you want to start the fire.
3. Leave to dry at room temperature.
4. Place on a fire resistant surface. Light a match, blow out the flame and touch the marked spot with the glowing tip (Figure 9.14). Watch your name go up in smoke!

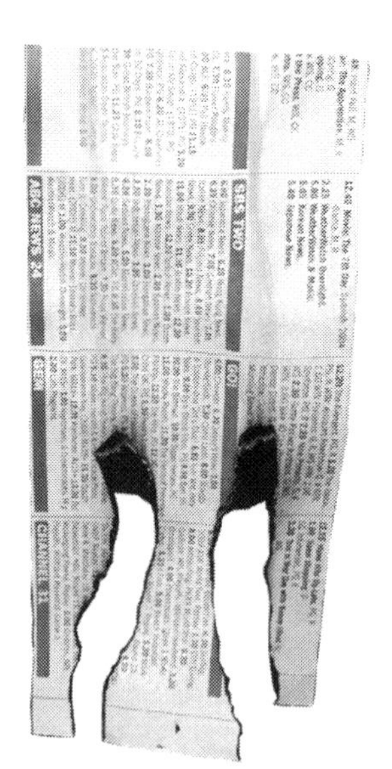

Figure 9.14
The Fire Race is on

A much faster combustion is attained with ***paper towel***. If you have access to a continuous roll of paper towel, use it for a **flaming poster** for the school open day. Give all a warm welcome by setting the words "Hot Chemistry!" alight.

In Figure 9.14 we started a "fire race" on a piece of newspaper. You can actually hear the fire racing forward (it sizzles) and see the tiny sparks around the combustion area due to the higher oxygen supply.

The simplest **fuse** I have ever prepared is made in a similar way. Wash thin packing string in hot water with soap. Rinse it well and dunk in the potassium nitrate solution for a few minutes, remove and let dry. Voila! This slow burning pyrotechnic fuse is only suitable for indoor use. To increase the combustion output simply take the rocket fuel mix (saltpetre/ sugar), prepare a paste with water and dip the washed string in the paste for a minute. Remove and leave to dry.

What is happening?

We have performed a very simple 'nitration' process of paper and string that is similar to the sugar-nitrate rocket system but different to the nitration process in guncotton and flash-paper preparation. In guncotton nitration processes new nitrate compounds form that are thermodynamically unstable:

$$3HNO_3 + C_6H_{10}O_5 \rightarrow C_6H_7(NO_2)_3O_5 + 3H_2O$$

Guncotton is a cellulose trinitrate which contains 12.5 to 13.5% nitrogen and is classified an explosive.

Our paper and string did not form cellulose nitrate. After the water has evaporated, the potassium nitrate was still present on the paper in a very finely divided state. The glowing match supplied the activation energy to get the combustion started and the decomposing potassium nitrate produced oxygen that sustained the combustion.

$$2KNO_3 \rightarrow 2KNO_2 + O_2$$

$$C_6H_{10}O_5 + O_2 \rightarrow CO_2 + H_2O + \text{energy}$$

Key Terms

Reaction surface area, reaction rate, activation energy, exothermic reactions, deflagration, decomposition, nitration processes

References

1. KN Sucrose propellant, Nakka, R., http://www.nakka-rocketry.net/sucrose.html Great website on experimental rocketry.
2. Stine G. H., Handbook of Model Rocketry, 7th edition, John Wiley & Sons, 2004
3. Have a look at the inspiring and challenging project named the "Sugar Shot to Space Program (SS2S)". The underlying goal of this program is to loft a rocket powered by "sugar propellant" into space, that is, 100 km (62 miles) above the earth's surface. Visit: www.sugarshot.org

Figure 9.15
Meat preservation - another application for saltpetre that is very exo-tasty

9 Touch Sensitive Chemicals

A house fly's demise!

In this chapter we will prepare crystals, so sensitive, a gentle touch with a feather or a breeze will activate them to decompose in a sharp explosion and a puff of purple gas!

Gunpowder never detonates when it is activated. It simply burns very fast, producing lots of gas to propel the projectile. But with this compound it is different. Here we have a shock wave traveling at a supersonic speed. This is a real explosive detonation!

The Background

Explosions are simply reactions in which a large amount of **energy** is released and a large **volume** of gaseous products are formed **very rapidly**.

Scientists have been aware for almost two centuries that the simple combination of ammonia gas and iodine crystals produces a solid that decomposes rather violently to nitrogen gas and iodine vapour.

This explosive has been the centre of some novel, stupid and plain dangerous applications. Some of you might have heard of mischievous school boys who booby trapped the school's floors by painting this compound on the floor. When dry, the teacher's presence is loudly announced down the passage when he/she steps on the crystals.

The production of explosives can never be described as completely safe, but I have used this method literally hundreds of times without any problem. If you adhere to the safety precautions and use sensible small quantities, then you will have the ability to demonstrate principles such as activation energy, potential chemical energy, exothermic energy, spontaneous processes and thermodynamic drivers without putting anyone in harm's way.

Preparing Ammonia Nitrogen Triiodide - a contact sensitive chemical[1]

The combination of iodine (I_2) and ammonia (NH_3) yields a brown solid which was thought, for many years, to be nitrogen iodide (NI_3). This simple reaction, however, does not produce pure NI_3 but rather two nitrogen iodide ammonia complexes that are extremely unstable in their dry state: [$NI_3.NH_3$] or [$NI_3.(NH_3)_3$]

$$3I_2\,(s) + 5NH_3\,(aq) \rightarrow NI_3.NH_3\,(s) + 3NH_4I\,(s)$$

The resulting $NI_3.NH_3$ solid is extremely unstable, touch and shock sensitive while NH_4I is a very stable compound. Even though $NI_3.NH_3$ can be triggered by a light breeze or a fly's footsteps when dry, it is not classified a high explosive such as secondary explosives used in military and blasting operations.

The decomposing detonation reaction can be presented by

$$8NI_3.NH_3\,(s) \rightarrow 5N_2\,(g) + 6NH_4I\,(s) + 9I_2\,(g) + \text{energy}$$

This is not an oxidation or combustion process as no oxygen is involved. It is a **decomposition redox reaction**. The loud, sharp detonation leaves an impressive plume of purple-brown iodine vapour.

History books tell us that two French chemists were actively pursuing the synthesis of these compounds:

- B. Courtois was the first to attempt the preparation of NI_3 in 1813;
- Another French chemist, Pierre Dulong lost two fingers and an eye in pursuit of another nitrogen trihalide, nitrogen trichloride NCl_3 , which forms explosive droplets. This is sometimes produced as co-product in wastewater treatment plants when chlorine reacts with nitrogen compounds.

---

[1] 'Contact sensitive chemicals' are also known as Touch Explosives, Explosive Paint or Wind Explosives.

Why is Ammonia Nitrogen Triiodide so extremely unstable?

There are two major *thermodynamic drivers* in the chemical decomposition reaction:

$$8NI_3.NH_3\ (s) \rightarrow 5N_2\ (g) + 6NH_4I\ (s) + 9I_2\ (g) + \text{energy}$$

Firstly. A major driver is the formation of products containing much less chemical potential energy than that of the reactants, leaving lots of energy to be released - a perfect exothermic reaction with a substantial **decrease in enthalpy**, ΔH^o. Note the large value of ΔH^o and small activation energy, E_a requirement in Figure 9.1.

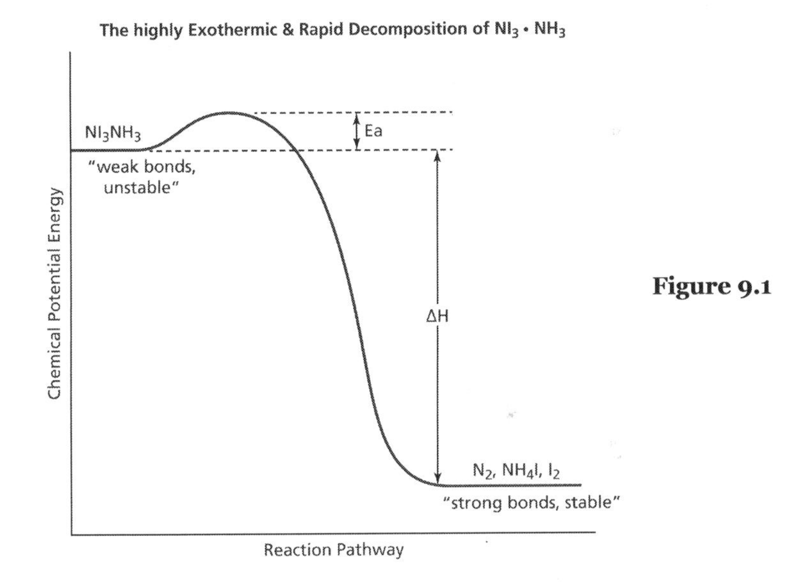

Figure 9.1

The products I_2 and N_2 both being pure elemental substances, present some of the lowest potential energy states, but it is the formation of nitrogen that is of much importance. Nitrogen is so stable because of its very strong covalent bonds. It takes lots of energy to break these bonds. The N_2 triple bond is one of the strongest chemical bonds known and prevents nitrogen gas from being chemically active, rendering it "inert" (Table 9.1).

Here is a thought: Even though nitrogen makes up the largest ratio of the earth's atmosphere (78%), it is not abundant in the earth's crust as it rarely reacts with other elements.
Nitrogen is energetically in a pretty comfortable and happy state!

		Bonding Energy kJmole^{-1}
Nitrogen	N_2	945
Hydrogen	H_2	436
Iodine	I_2	148

Table 9.1

The second thermodynamic driver in this spontaneous process is the huge **increase in entropy**: A gain of 14 moles in gaseous products guarantees a huge entropy increase (ΔS^o = pos.). Think: increase in randomness.

This reasoning fits in nicely with the principles of thermodynamics for chemical reactions. See Appendix A for more information.

Looking at commercial **explosives**, one finds that most contain high potential energy, **nitrogen compounds** with single or double bonds and have as a driving force the decomposition of these and the formation of the lower energy, more stable N_2 compound (ΔH^o = neg.). Because of these surplus energies, all explosives produce exothermic reactions. Furthermore. The increased entropy as more gaseous products form on decomposition, acts as another driver towards equilibrium and we come closer to understanding why explosives are as spontaneous and in some cases unstable as they are.

The following molecular structures of commercial and military explosives show the heavy reliance on "high energy" nitro groups. They all release substantial energy when forming stable N_2 molecules during decomposition.

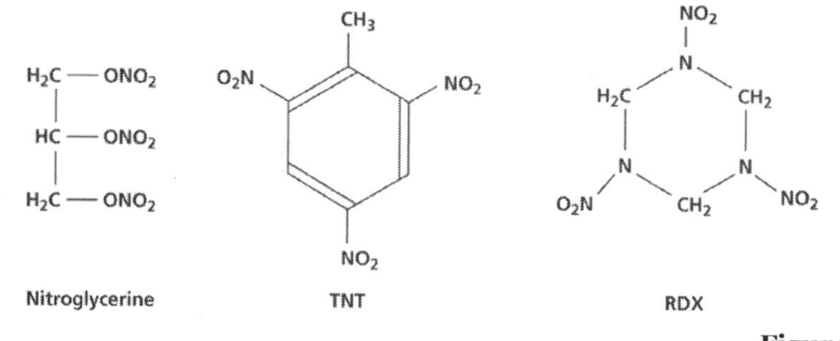

Figure 9.2

But, there is another important property, known as **kinetics**, that will determine if a thermodynamically unstable compound such as our triiodide, can earn the label of "explosive". As we all know, not all combustible spontaneous processes cause explosions, eg. the burning of paper or burning of a candle. The property required is that of a **high rate of chemical reaction**.

Study Table 9.2 and be prepared to be amazed:

Substance	Process	Internal energy MJ/kg	Combustion rate kJ/sec
Wood	burning	13	3.6
Coal	burning	28	7
Diesel	combustion	46	166
Gunpowder	deflagration	2	1,330,000
Nitroglycerin	detonation	4	20,000,000

Table 9.2

Nitroglycerin's impressive combustion rate creates a supersonic air pressure wave and it comes as no surprise that it has a detonation velocity around 7,000 m/sec ! So it is the **speed** (or **rate**) at which explosive chemicals combust or decompose that makes all the difference!

So much for the physical chemistry. Lets get practical.

Safety / Risk Assessment

[★★★★☆]

OK. Ammonia Nitrogen Triiodide is an explosive but there is no reason why a qualified science teacher with a chemistry background should not attempt to prepare at least a **small** quantity of this fascinating compound. The **wet** nitrogen triiodide complex is relatively safe to handle. However, if allowed to dry it becomes extremely unstable and even air movement can cause an unexpected detonation.

- ★ Never prepare more than is required. The substance cannot be stored
- ★ Only prepare small quantities at a time in an open, easily cleanable container
- ★ Do not move the explosives once positioned. Any disturbance can activate the explosive
- ★ Do not place in direct sunlight as spontaneous detonation will occur
- ★ Put a warning sign up and ensure that no one has access to the explosives area during the curing time
- ★ The detonation noise is loud - use ear protection and urge the audience to cover their ears

★ The detonation should only be performed behind a safety screen in a well ventilated area due to the formation of iodine vapours. A fume cupboard will be the first port of call
★ NEVER try to detonate with bare hands, even in small quantities. Use a long pole and feather (Figure 9.3)

Figure 9.3

Chemical Safety
★ Use standard safety procedures when handling all chemicals (protective glasses and gloves). Pour the ammonia in a fume cupboard
★ Chemicals should be locked away from students

Ammonium hydroxide (Ammonia solution) is an irritant with a strong pungent, suffocating odour. Only use in a well ventilated area.

- *Risk phrases: R36/37/38 Irritating to eyes, respiratory system and skin.*
- *Safety phrase: S2/3/26 Keep out of reach of children; Keep in a cool place; In case of contact with eyes, rinse immediately with plenty of water and seek medical advice.*

Iodine crystals are poisonous and produce a corrosive vapour. They are intensely irritating to eyes, skin and mucous membranes.

- *Risk phrases: R20/21 Harmful by inhalation and in contact with the skin.*
- *Safety phrase: S23/25 Do not breathe vapour; Avoid contact with eyes.*

What you will need

Chemicals

- **Iodine crystals**: I_2 (do not use Tincture of Iodine)
- **Ammonia** (aqueous solution): 15M NH_4OH, 25 to 35% solution
  You may use household Cloudy Ammonia which usually is an 4% ammonia solution but much better results are obtained with a "concentrated" ammonia solution.

Equipment

- Glass beaker, 100 ml
- Spatula
- Retort stand
- Ring clamp
- Masking tape
- Filter paper
- Feather duster or single feather taped to a stick (Fig. 9.3)
- Safety gloves
- Protective goggles & Ear muffs

Here's how

1. Transfer ¼ teaspoon iodine crystals to a small glass beaker.
   Careful: Iodine is an irritant.

2. Pour a small volume of ammonia solution over the crystals. Ensure that all of the iodine crystals are covered.
   Careful: Ammonia gas is an extreme irritant.

3. Swirl the beaker lightly for a few seconds and put aside for 5 minutes in a fume cupboard or well ventilated area. Brown solids will settle in the bottom. (If you use Cloudy Ammonia leave the solution aside for 15 minutes).

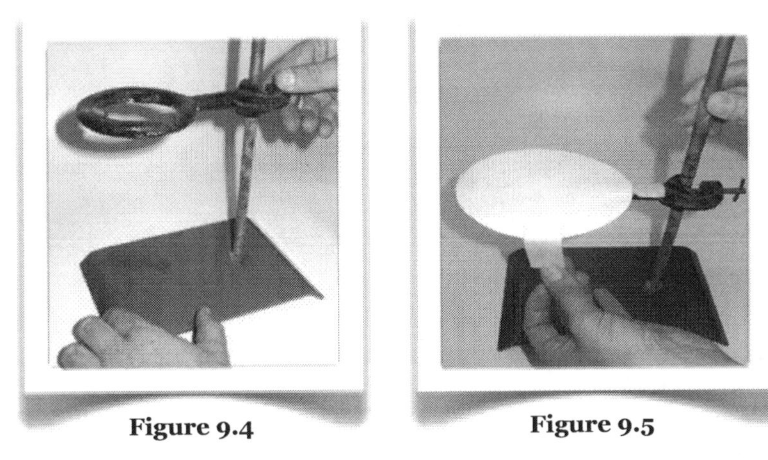

Figure 9.4 Figure 9.5

4. Prepare the retort stand with ring clamp (Fig. 9.4). Tape a double layer of filter paper to the ring using masking tape (Fig. 9.5). Place the stand where people (or pets when doing this at home) can't interfere. This has to be in a well ventilated area and definitely where there is NO wind or direct sunlight.

5. After 5 minutes - carefully pour all liquid in the glass beaker into a basin (decant), leaving the brown solid in the beaker (Fig. 9.6). Flush all liquids away with plenty of water.

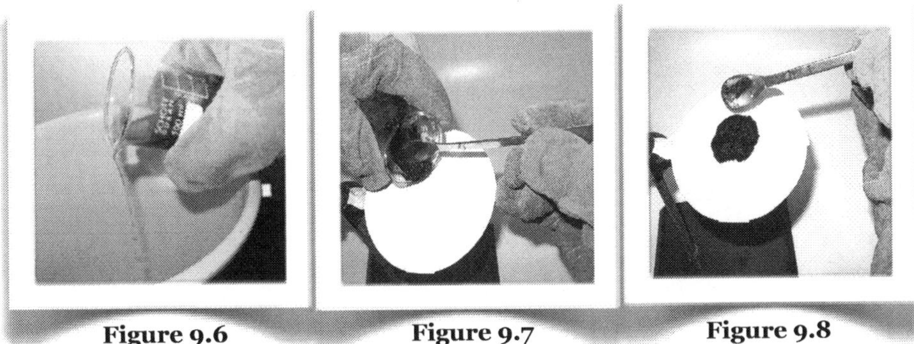

Figure 9.6 Figure 9.7 Figure 9.8

6. Transfer the brown solid to the filter paper with a spatula (Fig. 9.7). The solid compound is not yet touch sensitive but just in case . . . wear safety glasses.

7. Gently spread and smear the solids evenly over a small area of the paper. The paper will turn brown as it absorbs the brown liquid (Fig. 9.8). Wash the beaker & spatula thoroughly and wipe with paper towel before any residue dries.

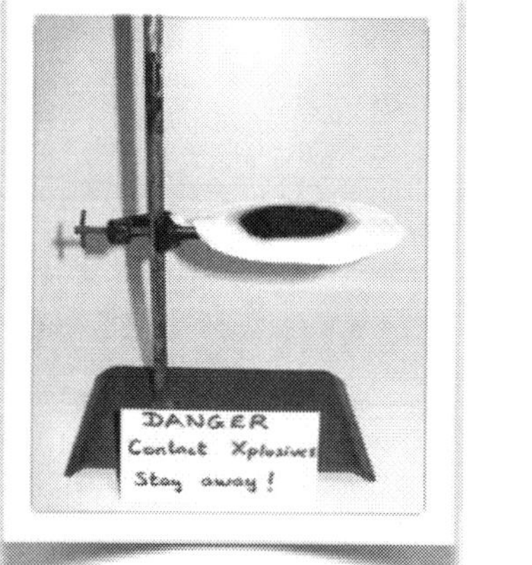

Figure 9.9

8. Place a sign next to the set-up warning people not to approach: " DANGER : CONTACT EXPLOSIVES. STAY AWAY". Check that no one can gain access to the area.

9. Leave UNDISTURBED to dry for 2 to 3 hours. This period is determined by many variables such as the quantity of solids, the humidity and ambient temperature. The ammonia nitrogen triiodide will be light grey in colour when dry (Fig. 9.9).

10. ACTION! *Wear eye and ear protection.*
Caution all bystanders to protect their ears.

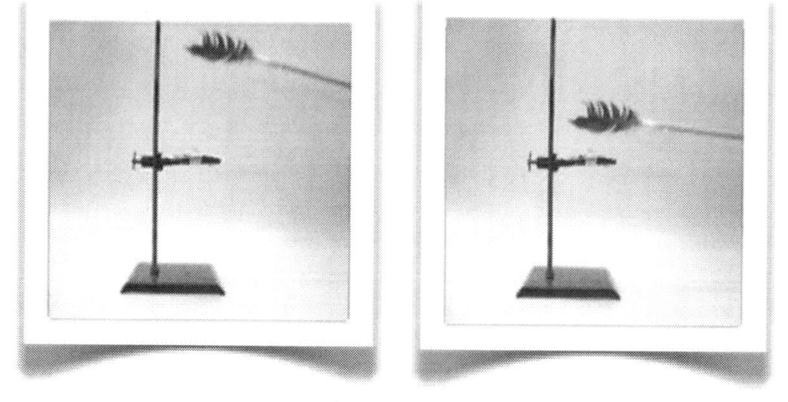

Figure 9.10

Figure 9.10

Figure 9.10: Approach the crystals with the feather mounted to the long pole and lightly, from a distance, dust the crystals with the feather . . .
BANG ! ! ! Immediately step away from the purple iodine vapour.

NOTE: If you do not get immediate detonation, leave the crystals for another 15 minutes to dry.

Teaching Extension

Here are two other fun activation methods:

Figure 9.11

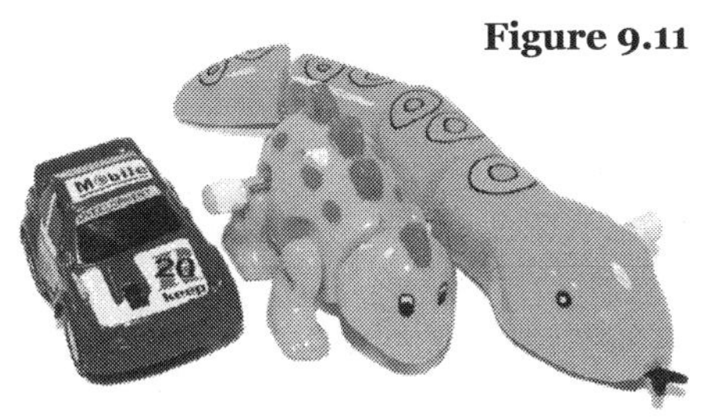

Buy a **wind-up toy** or **pull-back car** from a toy shop and make it walk a deadly destructive walk towards the touch sensitive triiodide. You may want to film this so it can be viewed in slow motion.

Figure 9.12

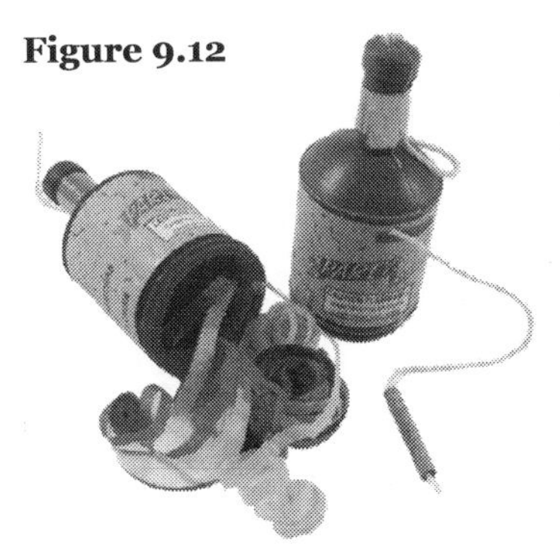

Party poppers are available for sale even in places where fireworks are banned. Their pyrotechnic parts can be very handy! Remove the bottom cardboard lid and paper pop-out parts. Cut the plastic body with scissors so you can pull the rod shaped 'detonator' free from the bottom. (Not from the top . . oops!) Lengthen the detonator's string with another piece and tape the detonator with masking tape to a sturdy surface. *This all still has to be done in a safe and well-ventilated area.* Place the wet nitrogen triiodide on top of the detonator to dry and

wait . . . do not detonate before the triiodide is dry.
If it malfunctions, wait another 15 minutes and set all strewn triiodide off with the feather duster.

Disposal & Clean-up

If detonation went well then you won't have much to discard. But if you set-off partially dry triiodide then the area may be "contaminated" with unreacted explosives. This is easily detected when walking around the area after the detonation. Sweep the area with a broom or duster.

Clean the direct explosion surface area with water and discard all paper residues in a waste bin. Any purple iodine stains will sublime within a few days. For short term results, remove the stains with a sodium thiosulphate solution.

Key Terms

Detonation vs deflagration, explosives, reaction kinetics, activation energy, spontaneous reactions, Gibbs free energy, enthalpy, entropy

References

1. Shakhashiri, B. Z.; Chemical Demonstrations: A handbook for teachers of chemistry, Vol. 1, The University of Wisconsin Press: Wisconsin, 1983; p 96 - 98

Appendix A:

Thermodynamics: The driving force behind chemical reactions

When chemists refer to a **spontaneous process** or reaction, it is one that occurs without the need for additional energy input after the reactants are mixed and the reaction is initiated.

Most chemical reactions in nature are exothermic and are usually spontaneous at 25°C, especially as many incorporate an increase in entropy (ΔH^o = neg. ΔS^o = pos.).

ΔG^o is the Gibbs free-energy change of a chemical reaction (after J W Gibbs who developed the concept in 1873). In chemistry it quantifies the maximum amount of energy available to perform useful work, from any chemical reaction. It allows one to determine whether a reaction will proceed and is minimized ($\Delta G = 0$) when a system reaches equilibrium at constant temperature and pressure.

The driving force for a chemical reaction depends upon two quantities:

1. The first is a **decrease in enthalpy**, ΔH^o, which represents the change in the internal potential energy of the atoms (breaking & making of bonds). On molecular level this means that there will be a tendency to form "strong" bonds at the expense of "weak" ones eg. the formation of N-triple bonds in explosive reactions.
2. The second is the drive toward an **increase in the entropy** of the system. An increase in the entropy (ΔS^o = pos.) favours a spontaneous reaction.

The combination of these two driving forces is represented as

$$\Delta G^o = \Delta H^o - T\Delta S^o$$

Together the terms enthalpy and entropy demonstrate that a system tends toward the lowest enthalpy and highest entropy, causing $\Delta G^o < 0$. Hence the rule of thumb:

"Every system seeks to achieve a minimum of free energy"

This is derived directly from the Second Law of Thermodynamics, which states that any physical and chemical change must result in an increase in the entropy of the universe.

Gibbs' free energy calculations allows one to determine whether a given reaction will be thermodynamically favorable (spontaneous). The sign of ΔG states that a reaction as written or its reverse process is the favorable step. If **ΔG is negative** then the **forward reaction is favoured** and energy will be released and vice versa for ΔG values that are calculated to be positive.

Almost all exothermic chemical reactions are spontaneous at 25°C and 1 atm. but endothermic reactions can be spontaneous too.
eg. The melting of ice at 25°C is a spontaneous endothermic phase change:

$$H_2O\ (s) \rightarrow H_2O\ (l) \qquad \Delta H^o = +6\ 000\ J/mole\ \text{(heat of fusion)}$$

Here the entropy increase is the only driving force in the reaction.

Other examples of spontaneous processes are the melting of ice, dissolution of salt in water, rusting of iron nails and the decay of wood buried in moist soil

Note: Spontaneous does not imply that the process occurs rapidly - iron can take a long time to corrode completely. The rate of a reaction has to do with the **kinetics** of a chemical reaction. There is no direct correlation between the rate of a reaction (kinetics) and the thermodynamic driving force as expressed by the free energy change.

Made in the USA
Lexington, KY
15 September 2011